Particle Science and Engineering
Proceedings of UK–China International Particle Technology Forum IV

Particle Science and Engineering
Proceedings of UK–China International Particle Technology Forum IV

Edited by

Xiaoshu Cai
University of Shanghai for Science & Technology, Shanghai, PR China
Email: usst_caixs@163.com

Jerry Heng
Dept of Chem Engineering, Imperial College London, London, UK
Email: jerry.heng@imperial.ac.uk

THE QUEEN'S AWARDS
FOR ENTERPRISE:
INTERNATIONAL TRADE
2013

The proceedings of the UK–China International Particle Technology Forum IV, held in Shanghai, China on 15-19 October, 2013.

Special Publication No. 347

Print ISBN: 978-1-84973-957-3
PDF eISBN: 978-1-78262-743-2

A catalogue record for this book is available from the British Library

Published by The Royal Society of Chemistry,
Thomas Graham House, Science Park, Milton Road,
Cambridge CB4 0WF, UK

Registered Charity Number 207890

Visit our website at www.rsc.org/books

Printed in the United Kingdom by CPI Group (UK) Ltd, Croydon, CR0 4YY, UK

Preface

The UK-China International Particle Technical Forum IV was successfully held between 14-17 October 2013 in Shanghai, China. It provided opportunities for scientists and engineers worldwide to discuss recent advances, to share their knowledge and to identify future research directions in the field of particle science and engineering and their roles in environment, energy, healthcare and other emerging applications. Nearly 100 scientists and researchers from universities, research institutions and industries in China, UK and other countries, e.g. Australia, Singapore, etc. participated in the conference. Prof Yulong Ding from University of Leeds, Prof. Chunzhong Li from East China University of Science and Technology, Prof. Mark Jones from University of Newcastle, Prof. Aibing Yu from University of New South Wales, Prof. Jin Y. Ooi from University of Edinburgh, and Prof. Jiyuan Tu from Tsinghua University were invited to have plenary lectures on their recent researches.

In the conference, a total of 97 papers were communicated, both oral and poster representations. The papers covered a wide range of topics which included synthesis and crystallisation, characterisation and measurement across length scales, multi-scale modelling and simulation, processing and handling of particulate system, nanoparticle technology and particle mechanics.

In this book, we present a selection of 18 papers related to the six topics mentioned above. The abstracts of all papers submitted to the conference and the presentations presented in the conference are attached in this book too. Among these papers, Peng et al. simulated the binary groups of particles with a sub-grid scale model in a riser, Zhang et al. studied the particle size influence on the turbulence characteristics within gas-solids pneumatic flows, and Wang et al. did three-dimensional simulation of the filtration process of polydisperse particulate matter by fibrous filter; Lv et al. investigated the droplet coalescence and oil-water separation characteristics of insulated electrode in electric dehydrator, Yang et al. discussed the coalescence and moving characteristics of droplets under pulsed DC electric field, and Fu et al. validated the powder properties measured by a rotational shear cell; Mao et al. presented the collection of nano-TiO_2 aerosol by using a novel wet sampler, Yang et al. described the numerical solution of dynamics of PM_{10} subjected to standing-wave acoustic filed, Zigan et al. described a design challenge for filters in the air extraction system, Li presented a comprehensive technology of particle characterization in one single platform, Xu et al. studied the squeeze flow of a bi-viscosity fluid between two rigid spheres. Cheong et al. discussed the discharge analysis of an industrial batch rotating drum, and Xu et al. prepared ^{3}-Al_2O_3 nano-particles by mechano- and sono-chemical reactions, and the other papers are related to parameter measurement of particle by imaging and ultrasonic method, which all reflected the hotspots in research and development in the field of particle science and engineering.

As editors, we are sincerely grateful to the authors and the reviewers for their excellent collaboration, and we would like to acknowledge Dr. Huinan Yang in University of Shanghai for Science and Technology for her efforts in publishing this book.

Editors

Prof. Dr. Xiaoshu Cai

University of Shanghai for Science & Technology

Dr. Jerry Y. Y. Heng

Imperial College London

Contents

List of Presentations

Plenary lecture

Name	Topic	Affiliation
Prof. Yulong Ding	Thermal energy storage materials - linking microstructures to device/system level performance	University of Leeds, UK; Chinese Academy of Sciences, China
Prof. Chunzhong Li	Chemical engineering fundamental for fabrication of functional nanomaterials - from structure control to process scale up	East China University of Science and Technology, China
Prof. Mark Jones	Recent developments in pneumatic conveying of fine powders	The University of Newcastle, Australia
Prof. Aibing Yu	Recent developments in particle scale modeling of particle-fluid flow	The University of New South Wales, Australia
Prof. Jin Y. Ooi	DEM modeling of cohesive and cementitious materials-verification, validation and applications	University of Edinburgh, UK
Prof. Jiyuan Tu	Micro/Nano Particle Transport and Deposition in Human Airways	Tsinghua University, China; RMIT University, Australia

Keynote talk

Measurement and validation of DEM simulations of ball motion within a planetary ball mill using Positron Emission Particle tracking (PEPT) M. Marigo The University of Birmingham
How impurity modifies crystal morphology: a structural investigation in sodium dithionate doped sodium chlorate crystal X. J. Lai, Z. P. Lan, K. J. Roberts, H. Klapper, L. P. Cardoso Institute of particle Science and Engineering, University of Leeds Institute of Crystallography, RWTH Aachen University Institute of Physics, University of Campinas
Nanoparticles for downstream separation of biopharmaceuticals J. Y.Y. Heng Imperial College London
Advances in particulate materials characterization using sorption probes D. R. Williams Imperial College
Nanomaterial characterization challenges P. Clarke Malvern Instruments Ltd

Ultrasonic process tomography applied to two-phase flow measurement   M. X. Su   University of Shanghai for Science and Technology
Oxidation and ignition characteristics of energetic nanomaterials   D.S. Wen   University of Leeds
Multi-scale particulate mechanics in multi-disciplinary engineering applications   S.J. Antony   University of Leeds
Bio-inspired biodegradable polymer nanoparticles for theranostic applications   R. J. Chen   Imperial College London
The application of computer modeling in pharmaceutical powder processing: an overview   C. Wu   University of Surrey
DNS and LES of particle flows in dense fluidized bed   K. Luo   Zhejiang University
Hybrid nanomaterials for electrochemical devices in energy management   P. S. Lee   Nanyang Technological University

Oral talk

Synthesized CaO Sorbents for CO_2 Capture   Y. Wang, J. Y.Y. Heng   Imperial College London, South Kensington Campus
Template-free and scalable synthesis of TiO_2 hollow spheres and its application in dye-sensitized solar cells   J.C. Huo, Y.J. Hu, C.Z. Li   East China University of Science & Technology
The effect of a and b sites substituting on humidity sensing properties of $BiTiO_3$-based power   R. Wang, D. Wang, X.J. Zheng   Xiangtan University   University of Shanghai for Science & Technology
Calcium borates nanostructures: hydrothermal formation, self-assembly, thermal conversion, and properties   W.C. Zhu, X. L. Wang, X. Zhang, Q. Zhang   Qufu Normal University   Tsinghua University
Room temperature NTC electrical switching based on MWCNTs/DI-water composites   R. T. Zheng   Beijing Normal University

Mass production of high purity ultrafine Ni powders in a fluidized bed reactor
J. Li, X. W. Liu, L. Zhou, Q. S. Zhu, H. Z. Li
Institute of Process Engineering, Chinese Academy of Sciences
Investigation of droplet coalescence and oil-water separation characteristics of insulated electrode in electric dehydrator
Y.L. Lv, Q. Zhang, L.M. He, X.M. Luo
China University of Petroleum
A numerical investigation of pressure and flow in a flat-bottomed model silo
Y. Wang, J.Y. Ooi
The University of Edinburgh
Advances in dust explosion protection technology
J.R. Taveau
Fike Corporation
Major improvement of CNTs' distribution in molten salts through adding MgO nanoparticles
G. Qiao, Y. L. Ding
University of Leeds
Validation of powder properties measured by a rotational shear cell
T. Freeman, X. Fu
Freeman Technology Ltd
Three-dimensional grapheme/CNTs/MnO2 ternary architectures for high-performance supercapacitors
Y. H. Dai, H. Jiang, Z. N. Deng and C.Z. Li
East China University of Science and Technology
Real-time nanoparticle sizing by image dynamic light scattering and the application
C. Z. Xu, X. S. Cai, J. Zhang
University of Shanghai for Science & Technology
Experimental study of gas-liquid two-phase flow in hydrophilic and hydrophobic surface
H. Cao, Y. Li, Y. Ding
University of Leeds
Institute of Process Engineering, Chinese Academy of Science
Hierarchical interconnected macro-/mesoporous Co-containing N-doped carbon for efficient oxygen reduction reactions
H.L. Jiang, Y.H. Zhu, X.L. Yang, C.Z. Li
East China University of Science and Technology
The latest progress in laser particle size measurement technology in china
Q. Y. Dong
Dandong Bettersize Instruments Ltd.
Parameter measurement system of Taylor flow in small channels
H. J. Li, H. F. Ji, C. Fu, B. L. Wang, Z. Y. Huang, H. Q. Li
Zhejiang University
The effect of modeled particle wall interaction on pneumatic conveying pressure drop prediction.
D. McGlinchey
Glasgow Caledonian University

Adhesion of single perfume-filled microcapsules on a model fabric surface investigated by AFM Y. He, J. Bowen, J. Smets and Z. Zhang University of Birmingham
Local gap and phase transition in the packing of uniform spheres Z. A. Tian, A. B. Yu University of New South Wales
The electrostatic charging trends of coal particles in a dense phase pneumatic conveying system F.F. Fu, C.L. Xu, S.M. Wang Southeast University
Comparative study of image analysis methods for particle mixing process in rotary drums X.Y. Liu, C.Y. Zhang, J.S. Zhan Hunan University Northeast University
Activities in ISO standardization of fine particle technologies R. L. Xu Micromeritics Instrument (Shanghai), Ltd.
Discharge analysis of an industrial batch rotating drum Y.S. Cheong, A. Zhao, H. Ahmadian, W. Bi, R. Shen Procter and Gamble
Mass transfer between bubbles and dense phase in gas fluidized beds H.Y. Xie Shanghai Second Polytechnic University
Silicon nanowire-supported bimetallic Cu@Ag nanocomposites synthesis and catalytic properties H. Zhong , X. L. Yang , Y. H. Zhu, C .Z. Li East China University of Science and Technology
Poorly soluble API-polymeric carrier interaction in spray dried powder K. Punčochová, F. Štěpánek, J. Heng Institute of Chemical Technology Prague Imperial College London, South Kensington Campus
Manufacture of functional ceramic nanopowders by thermal decomposition of metal-alginate gel structures Z. Wang, G. Kale, M. Ghadiri University of Leeds
Nanocarbon/sulfur composite as energy particles for high performance Li-S batteries J. Q. Huang, Q. Zhang, X. F. Liu, F. Wei Tsinghua University
Steady-state modeling of radial heterogeneity in the fully developed region of a CFB riser S.W. Hu, X.H. Liu, J.H. Li Institute of Process Engineering, Chinese Academy of Sciences
Coalescence and moving characteristics of droplets under pulsed DC electric field D.H. Yang, M.H. Xu, L.M. He, Y.L. Lü, H.P. Yan China University of Petroleum

A novel chemical synthesis of trivalent chromium microstructure Y. Bai Beijing Normal University
Erosion prediction of liquid-particle two-phase flow in pipe elbows via CFD-DEM coupling method Y. S. Wang State Key Laboratory of Multiphase Flow in Power Engineering, Xi'an Jiaotong University
Theoretical simulations for multiple charged particles travelling in an inductive charge sensor T. Hussain, T. Deng, D. I. Armour-Chelu and M.S.A. Bradley University of Greenwich
Development of an elastic-plastic force-displacement model for discrete element simulations D. M. Rathbone, M. Marigo, D. Dini, B. G. M. van Wachem Imperial College London
Two-stage modeling of the dispersion of a carrier-based dry powder inhalation system Z.B. Tong, R.Y. Yang, A.B. Yu, and H.K. Chan University of New South Wales
Three-dimensional simulation of the filtration process of polydisperse particulate matter by fibrous filter H. M. Wang, H. B. Zhao, C. G. Zheng Huazhong University of Science and Technology
Simulation of binary groups of particles with a sub-grid scale model in a riser W. G. Peng, F. L. Tian, S. L. Yan, Y. R. He Harbin Institute of Technology
New constitutive equations for solid stress in slightly wet particulate flow X. Yu, S. Generalis, R. Ocone and Y. Makkawi Aston University
A numerical investigation into the charge behavior of a spherical phase change material particle for high temperature thermal energy storage C. Li, Z. Sun, Y.L. Ding University of Leeds
The numerical simulation of bubbling fluidized beds with fine particles based on the modified agglomerate-force balance model Z. Zou, H. Z. Li, Q. S. Zhu Chinese Academy of Sciences
Multi-particle FEM modeling on microscopic behavior of 2D particle compaction Y. X. Zhang, X. Z. An, Y. L. Zhang Northeastern University
Comparison of condensation growth of insoluble and soluble particles under supersaturated vapor conditions G. S. Wen, F. X. Fan University of Shanghai for Science and Technology
Performance examination of a novel electrically-enhanced bag filter M. M. Yang, S. Chen, S. Q. Li, Q. Yao Tsinghua University

PREPARATION OF γ-Al$_2$O$_3$ NANOPARTICLES BY MECHANO-CHEMICAL AND SONOCHEMICAL REACTION

Bo Xu [*1], Shengjuan Li[2], Shulin Wang [2], Laiqiang Li[1]
[1] College of Energy and Power Engineering, University of Shanghai for Science and Technology, Shanghai 200093, P R China
[2] College of Material Science and Engineering, University of Shanghai for Science and Technology, Shanghai 200093, P R China

1   INTRODUCTION

The γ-Al$_2$O$_3$ nanosolids, a kind of porous activated alumina, are widely used in plastic, rubber, ceramics and fireproofing materials as the reinforcing agent, where its characteristics of high thermal stability, adhesion resistance, high mechanical strength and wear resistance are required. It has more remarkable advantages of enhancing ceramic in its compactness, finish, fracture toughness, resistance to creep and abradability of macromolecule materials. Besides, γ-Al$_2$O$_3$ is also a good dispersant, which can be uniformly dispersed in many solvents, e.g. water, ethanol, acetone, benzene and xylene etc [1-5].

In this paper, a novel method is applied to prepare porous γ-Al$_2$O$_3$ nanoparticles. At first, the 2h-milled aluminum powders are prepared as the starting material, then the powders react with water to produce Al(OH)$_3$ collosol in an ultrasonic water-bath, lastly, dehydrating, grinding, and roasting the Al(OH)$_3$ colloid at a given temperature to produce the porous γ-Al$_2$O$_3$ nanosolids. This method, which combines mechano-chemical and sonochemical reaction, is extremely innovative, and it may inspire new ideas for preparation of nanomaterials.

2   EXPERIMENTS

2.1 Preparation of aluminum ultra-fine particles

Experiment was conducted in a dry roller vibration mill (RVM) at room temperature. The RVM has a chamber of 2.5L, equipped with a motor of 0.12kW[6,7]. To provide proper atmosphere and prevent dust explosion, the entire operation was performed in a glove box filled with argon. In a typical experiment, 100g raw aluminum powder (with average sizes of 150μm, purity higher than 99.5%, purchased from Guoyao group chemical reagent Co. Ltd., China) is placed in the grinding chamber with the stainless steel rods as grinding medium, and the weight ratio between the medium and powder is 60:1. The 2h-milled sample is then collected for later use.

2.2 Synthesis of γ-Al₂O₃ porous nanoparticles

2.0g of 2h-milled aluminum powder was put into a beaker with the water of 50mL The beaker was placed in an ultrasonic water-bath (DL-120J, made in China), setting the supersonic frequency of 100kHz at the room temperature. Under the high temperature and high pressure circumstances generated by the ultrasonic cavitation, the energy stored in the material was fully released, and the particles reacted with water to produce Al(OH)₃ white latex. In succession, it was dehydrated in a drying cabinet (101A-1, made in China) at 80℃ for 6h to remove the excess water. Lastly, grind the gel into white powder and place it in an electrical resistance furnace (SX2-5, made in China) to calcined 4h at 160℃, the porous γ-Al₂O₃ nanoparticles were obtained. The experimental flow chart is shown in Figure 1.

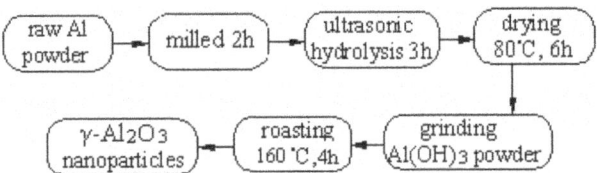

Figure 1 *Experimental procedure*

2.3 Characterizations of the particles

Structural phase analysis was carried out with D/max-γA X-ray diffractometer (XRD), using Ni-filtered Cu–Kα radiation as the X-ray source. The scanning speed was $4°$/min. The morphology of the sample was observed using a FEI Quanta 450 scanning electron microscope (SEM), and the accelerating voltage was 20 KV. Transmission electron microscopy (TEM) image was recorded on Tecnai G² 20 S-Twin electron microscopy at 200 kV. All measurements were carried out at room temperature.

3   RESULTS AND DISCUSSION

3.1 The structure analysis of aluminum powders

From the SEM image (Figure 2a), we can see that the particles sizes are in the range of 0.5-0.8μm after milled for 2h, the size is in good agreement with the particle size measured from TEM image in Figure 2b, and the SEM image shows a uniform, flaky crystal pattern

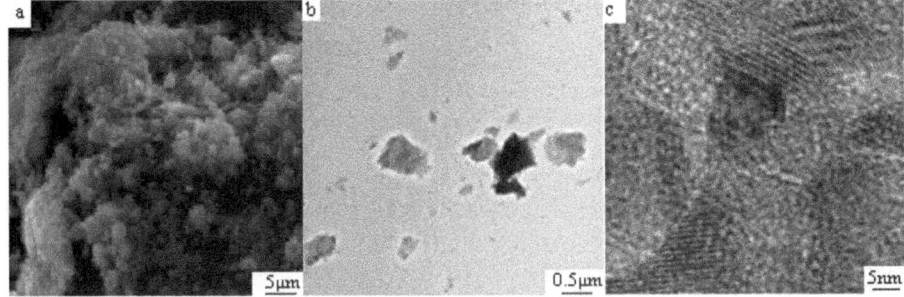

Figure 2 *SEM, TEM and HRTEM images of the powder milled 2h*
a. SEM image   b. TEM image c.HRTEM image

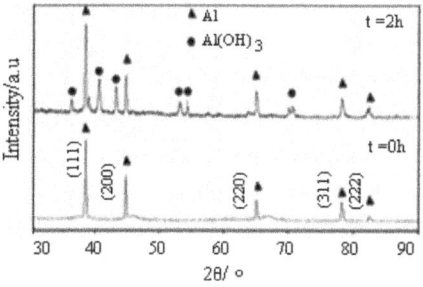

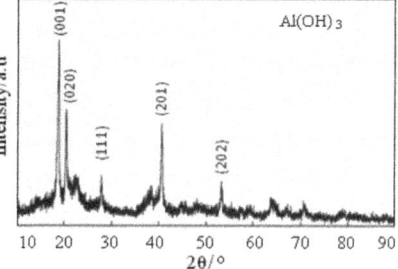

Figure 3 *XRD patterns of Al powder*   **Figure 4** *XRD pattern of Al(OH)₃*

with 0.5μm in width and 0.8μm in length. The microstructure of the nanoparticle is also confirmed by HRTEM (Figure 2c). The result shows that the solid particles, under the action of mechanical force, generate a mass of deformation and dislocation flaws, leading to the material to a metastable high-energy state, which is favorable for mechano-chemical reaction[8].

Figure 3 illustrates the XRD patterns of the raw and the 2h-milled aluminum powders respectively. In Figure 3, after two hours of milling, the diffraction peaks are still indexed to the face-centered cubic lattice aluminum, but $Al(OH)_3$ should be recognized as a minor phase. This is because the material is in a metastable, high-energy activity state during the grinding process, which has a high mechano-chemical activity, therefore, as the particles are exposed in atmosphere, the solid particles will react with the water vapor, and then a small amount of $Al(OH)_3$ is generated.

3.2 Ultrasonic hydrolysis reaction

The ultrasonic hydrolysis reaction is the next step to create the γ-Al_2O_3. The hydrolysis experiment was completed in an ultrasonic water-bath. Under the local high temperature and high pressure circumstance generated by cavitation effect, the water is decomposed into the ·H and ·OH free radicals. The O_2 dissolved in the solution is also a decomposition response and produces ·O free radical. Because the ·OH possesses unpaired electrons, it has a strong oxidability, so it is prone to induce the redox reaction. The reaction process is

$$H_2O \rightarrow ·OH + ·H$$
$$O_2 \rightarrow ·O + ·O$$
$$O + H_2O \rightarrow ·OH + ·OH$$
$$O_2 + ·H \rightarrow ·OH + ·O$$
$$OH + ·OH \rightarrow H_2O_2$$
$$Al + H_2O_2 + ·H \rightarrow Al^{3+} + OH^- + H_2O$$

The cavitation bubbles, in the collapse process, raise the liquid foam-core temperature up to 5200K and the pressure to 5.05×10^7 Pa. The local high temperature and high pressure condition exists only a short time, causing the temperature gradient as high as 109K/s, and at the same time, accompanied by a strong shock wave and micro-jet with the speed up to 400km/h, which provides an extreme physical and chemical condition for such difficult or even impossible chemical reaction[9-11].

On the other hand, in the milling process, the specific surface area of the solid particle increases as the particle size decreases. The adsorptivity of the particle surface is strengthened, and the surface free energy will change the electric charging into free radical, exoelectron emissing etc, so the chemical reaction rate is improved greatly [12-14]. Under the loading of mechanical force, the solid particles have lattice distortion, dislocation and other defects, which will also change the equilibrium of the chemical reaction and the activation energy, thus obtain an extremely high chemical reaction activity[15]. In the local thermal disturbance and tension stress field generated by ultrasonic cavitation, the particles fully release the energy stored in the material, and reacted with the water to produce $Al(OH)_3$ nanoparticles in a short period of time, as the XRD pattern shown in Figure 4.

3.3 γ-Al_2O_3 nanoparticles characteristics

As shown in Figure 5, it is clear that $Al(OH)_3$ is transformed into γ-Al_2O_3 after dried and roasted. From the TEM image (Figure 5a), it can be seen that the products are of porous, flaky nanostructures, no obvious aggregation, and the average particles sizes are in the range of 30-50nm. Figure 5b shows the XRD diffraction pattern of the sample. The diffraction peak position is consistent with the one in XRD standard diffraction card of γ-Al_2O_3.

4   CONCLUSIONS

After the vibration milling for 2 hours, the raw aluminum powder generates a large number of strain and dislocation defects. The material is in a metastable, high-energy active state. In the local thermal disturbance and tension stress field generated by the ultrasonic cavitation, the particles react with the water to produce white Al (OH) ₃ sol in a short time. The porous γ-Al_2O_3 nanoparticles are successfully obtained by drying, grinding and roasting the Al (OH) ₃. Compared with other preparation methods, this novel method can dramatically reduce the production cost.

ACKNOWLEDGEMENT

This project is supported by Shanghai Committee of Science and Technology, China (No.1052nm02900) and University of Shanghai for Science and Technology, China (No. 1D-11-301-002).

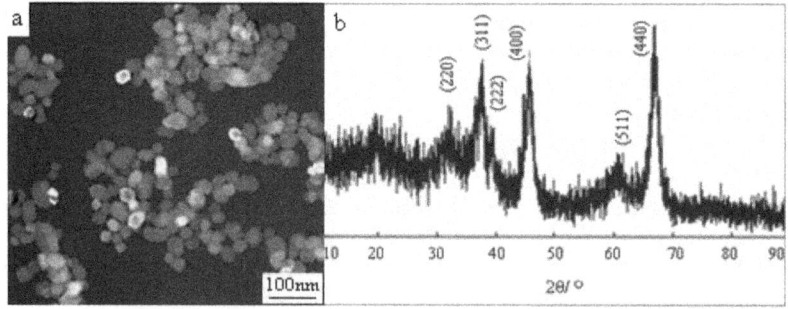

Figure 5 *TEM image and XRD pattern of γ-Al₂O₃*

a. TEM image of γ-Al₂O₃ b. XRD pattern of γ-Al₂O₃

References

1. Morajkar PP, Fernandes JB. Ceram Int 2010; 11: 414-18.
2. Ao Y, Yang Y, Yuan S and Hu H. Ceram Int 2007; 33: 1547-51.
3. Noguchi T, Matsui K, Islam NM, Hakuta Y and Hayashi H. J Supercrit Fluid 2008; 46: 129-35.
4. Liu XL. Mater Sci Eng B 2010; 175: 86-9.
5. Yang HM, Liu MZ and Ouyang J. Appl Clay Sci 2010; 47: 438-43.
6. Wang SL. Progress in Natural Science 2002; 2: 336-41.
7. Wang SL, Li SJ, Du YC, Xu B, et al. Progress in Natural Science 2006; 4: 441-44.
8. Xu B, Wang SL, Li SJ and Li LQ. Acta Phys Sin 2012; 3: 030703-1-5.
9. Hoffmann MR, Hua I and Hochemer R. Ultrason Sonochem 1996; 3: 163-68.
10. Bogdan N. Ultrason Sonochem 2007; 14: 13-7.
11. Xu B, Wang SL, Li LQ and Li SJ. Acta Phys Sin 2012; 9: 090201-1-5.
12. Kostic E, Kiss S, Boskovic S and Zec S. Powder Technol 1997; 91: 49-53.
13. Yuan TC, Cao QY and Li J. Hydrometallurgy 2010; 104: 136-40.
14. Kumar S, Kumar R. Ceram Int 2011; 37: 533-39.
15. Tkácová K, Heegn H and Stevulová N. Int J Miner Processing 1993; 40:17-31.

ANALYSIS OF SQUEEZE FLOW OF A BI-VISCOSITY FLUID BETWEEN TWO RIGID SPHERES

C H Xu, M Zhang, Y Xu and L N Zhang

College of Science, China Agricultural University, Beijing, 100083

1 INTRODUCTION

Recently, a science named granular matter mechanics which studies the balance and the characteristics of motion as well as the application of the granular matter is formed[1]. The study of granular matter mechanics aims mainly at dry particle system. If the particles have wet surfaces, the liquid bridges form at the contact point when the spheres approach each other. Meanwhile, complicated interconnected structure and flow law which influences directly the deformation and strength of the particle are created. At present, one of the difficulties is the squeeze flow and tangential movement of an interstitial liquid between particles [2-9].

The use of Bingham fluid model in the problem of squeeze flow results in no relative movement between two spheres or parallel disks, which named yield-surface paradox. By now the fluid model is too ideal to describe the physical nature. Based on general bi-viscosity fluid model, Zhu et al. created the theoretical analysis of the influence of electric field strength to the squeeze flow between disks, which solved the problem of yield-surface paradox. In terms of squeeze flow between two spheres, Adams and Edmondson [10] formulated the lubrication problem for a power-law fluid between two equal spheres. Lian [11] studied the squeeze force and liquid bridge force of a Newtonian fluid between two spheres which combined with the simulation of DEM. Rodin [12] considered near touching unequal spheres embedded in a power-law fluid and obtained the squeeze force. The group of the China Agriculture University studied deeply normal and tangential movement of an interstitial liquid between two spheres. They obtained analytical solution of squeeze viscous force as well as the tangential force and moment, which furthered the development of this difficult problem [13-18].

2 MATHEMATICAL MODEL

Consider two rigid spheres S_1 and S_2 of radii R_1 and R_2, translating towards each other with a relative velocity V_Z along axis Z and the bi-viscosity fluid between two spheres is squeezed, as shown as Figure 1, where B is the radius of liquid bridge, which is equal to R^* in the immersed system and s_0 is the minimum gap between two spheres, which make $s_0 \ll \min(R_1, R_2)$ satisfied. In the narrow gap, the two near surfaces of the spheres can be approximated by the following expressions:

$$\begin{cases} z_1(r) = s_0 + \dfrac{r^2}{2R_1} & (S_1) \\ z_2(r) = -\dfrac{r^2}{2R_2} & (S_2) \end{cases} \tag{2.1}$$

where r is the radial distance.

The separation distance of the spherical surface is represented as:

$$s(r) = s_0 + \frac{r^2}{2R^*} \tag{2.2}$$

where R^* is the harmonic radius, which is defined as: $2(R^*)^{-1} = (R_1)^{-1} + (R_2)^{-1}$

The boundary conditions between the spherical surfaces may be expressed as:

$$\begin{cases} v_r = 0; \quad v_z = -V_z & (S_1) \\ v_r = 0; \quad v_z = 0 & (S_2) \end{cases} \tag{2.3}$$

The constitutive relation of a bi-viscosity fluid is given as:

$$\begin{cases} \sigma_{rz} = \eta_r \dfrac{\partial V_r}{\partial z} & (|\sigma_{rz}| < \sigma_l) \\ \sigma_{rz} = sign(\dfrac{\partial V_r}{\partial z})\sigma_0 + \eta \dfrac{\partial V_r}{\partial z} & (|\sigma_{rz}| \geq \sigma_l) \end{cases} \tag{2.4}$$

where σ_{rz} is shear stress, σ_l is dynamic yield stress, σ_0 is yield stress; η_r and η is viscosity. When $|\sigma_{rz}| \geq \sigma_l$, the fluid flows by the viscosity η, and when $|\sigma_{rz}| < \sigma_l$, the

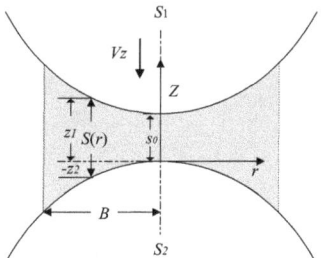

Figure 1. *Schematic diagram of the fluid and the squeeze flow between two rigid spheres.*

viscosity η_r is larger. The viscosity ratio is equal to η/η_r, which is generally equal to 10^{-5} to 10^{-2}. The boundary of the two regions is yield surface[6]. Obviously, the relation of σ_1 and σ_0 can be expressed as:

$$\sigma_0 = \sigma_1(1-\gamma) \tag{2.5}$$

3. THEORETICAL ANALYSIS

3.1 The determination of half thickness of unyielding region

The continuity equation for an incompressible fluid becomes:

$$\frac{1}{r}\frac{\partial}{\partial r}(rv_r) + \frac{dv_z}{dz} = 0 \tag{3.1}$$

The term for pressure gradient in z direction is negligible, so the momentum can now be reduced to the following expression:

$$\frac{dp}{dr} = \frac{\partial \sigma_{rz}}{\partial z} \tag{3.2}$$

Integrating momentum Equation (3.2) obtains:

$$\sigma_{rz} = \frac{dp}{dr}(z - z_0) \tag{3.3}$$

where $z_0 = \frac{z_1 + z_2}{2}$ is defined as the location where the shear stress equals zero. One can define $|\sigma_{rz}|$ equals σ_1, the position of yield surface can be expressed:

$$\begin{cases} z_{\sigma_1} = z_0 + t & (z_0 \leq z \leq z_1) \\ z_{\sigma_2} = z_0 - t & (z_2 \leq z \leq z_0) \end{cases} \tag{3.4}$$

where $t = \sigma_1 \left(-\dfrac{dp}{dr}\right)^{-1}$.

Substituting Equation (2.4) into Equation (3.2) and upon integration, the radial velocity V_r can be expressed as:

$$V_r = \begin{cases} \dfrac{1}{2\eta}\dfrac{dp}{dr}\left[(z-z_0)^2 - \dfrac{s^2}{4}\right] + \dfrac{1}{\eta}\sigma_0(z-z_1) & (z_{\sigma_2} \leq z \leq z_1) \\[2ex] \dfrac{1}{2\eta_r}\dfrac{dp}{dr}(z-z_0)^2 + \dfrac{1}{8\eta}\dfrac{dp}{dr}(4t^2 - s^2) - \dfrac{1}{2\eta}\sigma_0(s-2t) - \dfrac{1}{2\eta_r}\dfrac{dp}{dr}t^2 & (z_{\sigma_2} \leq z \leq z_{\sigma_1}) \\[2ex] \dfrac{1}{2\eta}\dfrac{dp}{dr}\left[(z-z_0)^2 - \dfrac{s^2}{4}\right] - \dfrac{1}{\eta}\sigma_0(z-z_2) & (z_2 \leq z \leq z_{\sigma_2}) \end{cases} \tag{3.5}$$

Integrating continuity equation (3.1) combined with boundary condition one gets:

$$\int_{z_2}^{z_1} V_r dz = \frac{1}{2} r V_z \tag{3.6}$$

Substituting Equation (3.5) into Equation (3.6) we gets the expression of t:

$$t^3 - \frac{3}{8\gamma + 4}\left(s^2 + 2\frac{\eta r V_z}{\sigma_0}\right)t + \frac{1}{8\gamma + 4}s^3 = 0 \tag{3.7}$$

Obviously, t is related to the pressure distribution gradient, we can obtain the dimensionless equation:

$$\bar{t}^3 - \left(\frac{3}{8\gamma + 4} + \frac{6d}{8\gamma + 4}\frac{\bar{r}}{\bar{s}^2}\right)\bar{t} + \frac{1}{8\gamma + 4} = 0 \tag{3.8}$$

where $\bar{t} = \dfrac{t}{s}$, $\bar{r} = \dfrac{r}{B}$, $d = \dfrac{\eta B V_z}{\sigma_0 s_0^2}$, $\bar{s} = \dfrac{s}{s_0} = 1 + c\bar{r}^2$, $c = \dfrac{B^2}{2R^* s_0}$. Parameter c reflects the change

of distance of the two spheres. The value of η and σ_0 is certain for any fluid. So when B and s_0 is certain, we can consider d as the nominal speed [14].

Equation (3.8) is the classical Cardan equation and a solution to Equation (3.8) is obtained as:

$$\bar{t} = 2\xi^{\frac{1}{3}}\cos\left(\Phi + \frac{4\pi}{3}\right) \tag{3.9}$$

where $\xi = \left(\dfrac{1}{8\gamma + 4} + \dfrac{2}{8\gamma + 4}\dfrac{d\bar{r}}{\bar{s}^2}\right)^{\frac{3}{2}}$, $\Phi = \dfrac{1}{3}\arccos\left(-\dfrac{1}{2(8\gamma + 4)\xi}\right)$.

The solution of Equation (14) is expressed as:

$$\bar{t} = \frac{1}{2}\left[-\bar{t_1} + \sqrt{(\bar{t_1})^2 + (\bar{t_1})^{-1}}\right] (0 \le \bar{t} \le 0.5) \tag{3.10}$$

where $\bar{t_1} = \dfrac{1}{2}\left[\left(-1 + \sqrt{1 - 64\xi^2}\right)^{\frac{1}{3}} + \left(-1 - \sqrt{1 - 64\xi^2}\right)^{\frac{1}{3}}\right]$

3.2 Pressure Distribution and Squeeze Force:

According to Equation (3.10), we can get the expression of the location of the yielding region as:

$$z_{\sigma_0} = z_0 - sign(z - z_0)\bar{t}s \tag{3.11}$$

Once the thickness of the unyielding region is determined, its relation with the pressure distribution can be obtained as:

$$\frac{d\,p}{d\,r} = -\frac{\sigma_0}{t} = -\frac{\sigma_0}{s\,\bar{t}} \tag{3.12}$$

Integrating Equation (3.12), one can get the pressure distribution:

$$p(\bar{r}) = \int_r^B \frac{dp}{dr}\,dr = -\frac{\sigma_0 B}{s_0}\int_{\bar{r}}^1 \frac{1}{s\,\bar{t}}\,d\bar{r} \tag{3.13}$$

Then the dimensionless pressure is defined, so the above equation becomes:

$$\bar{p}(\bar{r}) = \frac{p(\bar{r})}{p_0} = \int_{\bar{r}}^1 \frac{1}{s\,\bar{t}}\,d\bar{r} \tag{3.14}$$

where $\quad p_0 = -\dfrac{\sigma_0 B}{s_0}$.

Integrating Equation (3.14) , the squeeze force can be obtained as:

$$F = 2\pi\int_0^B p(r)r\,dr = \frac{\pi\sigma_0 B^3}{s_0}\int_0^1 \frac{\bar{r}^2}{s\,\bar{t}}\,d\bar{r} \tag{3.15}$$

Then we can get the following dimensionless form:

$$f = \frac{F}{F_0} = \int_0^1 \frac{\bar{r}^2}{s\,\bar{t}}\,d\bar{r} \tag{3.16}$$

in which $\quad F_0 = \dfrac{\pi\sigma_0 B^3}{s_0}$.

4. NUMERICAL FITTING AND ANALYSIS

In order to characterize the squeeze flow of the bi-viscosity fluid, numerical results for the thickness of the yielding region as well as the pressure distribution and squeeze force are calculated by own calculation program, and then the change laws are discussed in conditions of selecting different parameters.

4.1 Numerical analysis of pressure distribution

The plots of dimensionless of half thickness of unyielding region profiles for different viscosity ratio [3]. From Figure 2 it is clear that the range of yielding region increases as viscosity ratio reduces. The location of the yielding region changes little and tends towards stability as [3] is smaller than 10^{-3}. Therefore the viscosity ratio is defined as 10^{-3}.

From Figure 3 it is seen that the dimensionless half thickness of unyielding region $\bar{t}$ reduces as d increases, because c is fixed. So it demonstrates that larger nominal velocity will make more fluid yield. For the parameter d, the parameters of fluidal physical characteristic η and σ_0 are dominant factors as B and s and V_z are fixed. For a fixed η is larger or a fixed σ_0 is smaller, the range of yielding region increases as the range of unyielding region reduces, and vice versa. From Figure 4 it is seen that the dimensionless half thickness of unyielding region $\bar{t}$ increases as c increases, because d is fixed. It seems that the smaller the distance of the two spheres is, the larger the range of unyielding region

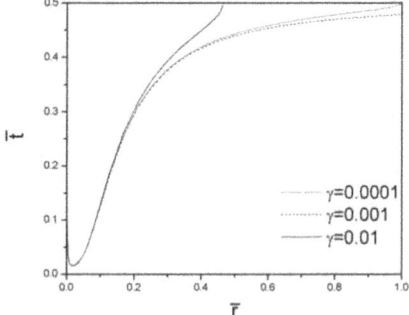

Figure 2. *Dimensionless half thickness of unyielding region $\bar{t}$, different values of viscosity ratio* γ *(c=1000,d=1000)*

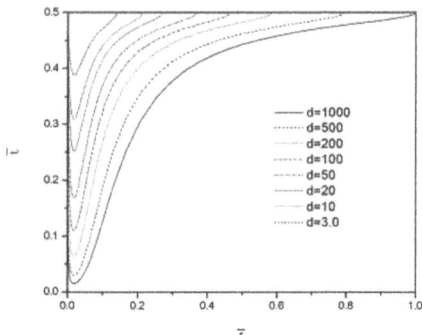

Figure 3. *Dimensionless half thickness of unyielding region $\bar{t}$, different values of parameter d (c=1000)*

is and the smaller the range of yielding region is, and vice versa.

The plots of the pressure distribution for a number of d and c parameters are shown in Figure 5 and Figure 6. From Figure 5, it shows that the dimensionless pressure reduces as parameter c increases with a fixed parameter d. It means that the smaller the distance between two spheres is, the smaller the dimensionless pressure is. The change trend of the extension of the unyielding region which shows in Figure 4 is the same as that the change trend of the pressure which shows in Figure 5. From Figure 6, it shows that the dimensionless pressure increases as the parameter d increases with a fixed parameter c, which means that the pressure increases as the nominal velocity increases. The change trend of the extension of the unyielding region which shows in Figure 3 is the same as that the change trend of the pressure which shows in Figure 6. Moreover, From these Figures it can be seen that the pressure increases rapidly as $\bar{r}$ increases, with the maximum values always being located at the center of contact. The change trend is the same as the power-law and Bingham fluid.

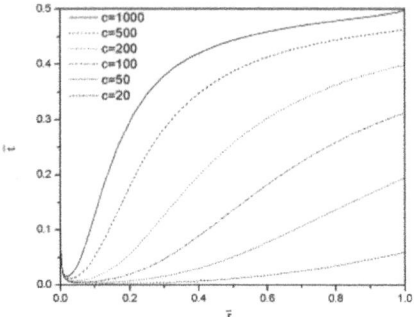

Figure 4. *Dimensionless half thickness of unyielding region $\bar{t}$, different values of parameter c (d =1000)*

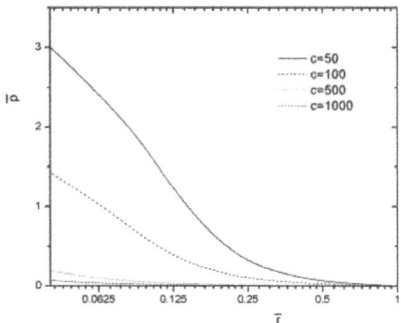

Figure 5. *Dimensionless pressure distribution $\bar{p}(\bar{r})$, different values of parameter c (d =100)*

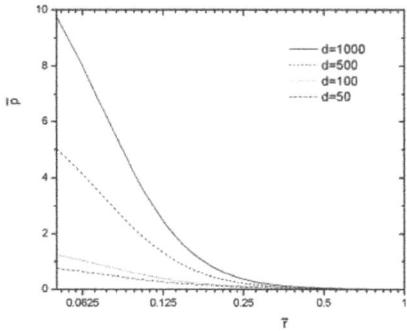

Figure 6. *Dimensionless pressure distribution $\bar{p}(\bar{r})$, different values of parameter d (c =100)*

4.2 Numerical analysis of squeeze force

The plots of the squeeze force for a number of *d* and *c* parameters are shown in Figure 7 and Figure 8. From Figure 7, it shows that the squeeze force reduces as parameter *c* increases with a fixed parameter *d*. It means that the smaller the distance between two

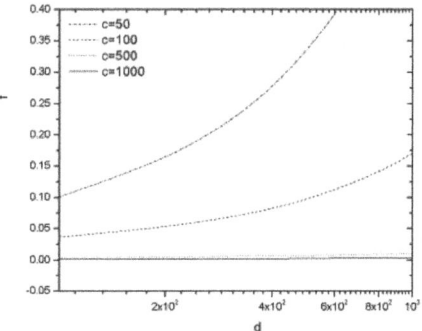

Figure 7. *Dimensionless squeeze force f, different values of parameter c*

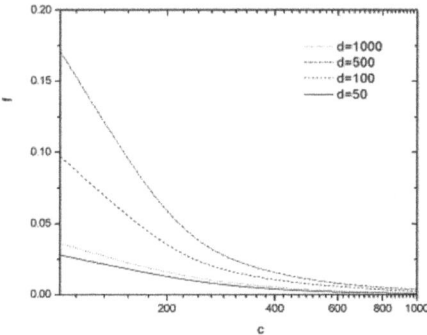

Figure 8. *Dimensionless squeeze force f, different values of parameter d*

spheres is, the smaller the squeeze force is. The change trend of squeeze force which shows in Figure 7 is the same as that the change trend of the pressure which shows in Figure 5. From Figure 8, it shows that the dimensionless pressure increases as the parameter *d* increases with a fixed parameter *c*, which means that the squeeze force increases as the nominal velocity increases. The change trend of squeeze force which shows in Figure 8 is the same as that the change trend of the pressure which shows in Figure 6.

5   CONCLUSIONS

Based on the *Reynolds* lubrication theory, from the analysis of squeeze flow of an interstitial bi-viscosity fluid between two rigid spheres, the following conclusions were drawn.

1.   According to the characteristics of stress distribution, the fluid was divided into both yielding and unyielding regions. Along with the variation of parameters, differences in the thickness of the two regions were discussed and the pressure distribution in these regions was analyzed. The numerical curve whose position was decided by the values of parameter *c* and *d* of the half thickness of unyielding region was the boundary of unyielding and yielding regions.

2. The integral expression of pressure distribution and viscous force were derived. According to the numerical solution, the change law of the pressure distribution and the viscous force were analyzed along with the variation of parameters c and d.

3. Based on the *Reynolds* lubrication theory, approximate solutions were obtained due to the simplification of equations during derivation. So the range of use of the solution need be proved further by experiments.

ACKNOWLEDGEMENT

The authors acknowledge the support from the National Natural Science Foundation of China (NSFC, Grant No. 11272341)

References

1   Q. C. Sun, G. Q. Wang,. Intro. Mech. *Granular Matter*, ,Beijing: Science Press, 2009 ,ch.1,p.4 (in Chinese)

2   G. X. Li,   *Advance Soil Mechanics*, Beijing: Tsinghua University Press, 2004, ch.1,p.3 (in Chinese)

3   D. G. Fredlund, H.Rahardjo, *Soil Mech. unsaturated soils*, New York: John Wiley & Sons, Inc,1993,ch. 2,p.52

4   D. N. Smyrnaios, J.A.Tsamopoulos, *J. Nonnewton Fluid Mech.* 100 (2001) 165-190.

5   J. D. Sherwood, D.Durban, *J. Non-Newtonian Fluid Mech.* 1998, 77,115.

6   K. Q. Zhu, R.Ge, B.S. Xi,. *J. Tsinghua Univ. (Sci. Tech.)* 1999, 39, 80 (in Chinese)

7   H. M. Laun, M.Rady, O.Hassager, *J. Non-Newtonian Fluid Mech.* 1999, 81,1

8   C. H. Xu, W.B. Huang,Y. Xu, *J. Transactions of CSAE.* 18(5)(2002) 19-22.

9   C. H. Xu, W.Yang, *J. China Agriculture Univ. (Sci. Tech.).* 2009, 14,27. (in chinese)

10 M. J. Adams, B.Edmondson, In: *Tribology in Particulate Technology,* ed. Briscore B J & Adams M J., 1987, 154.

11 G. P. Lian, C.Thornton, M.J.Adams, *J. Colloid and Interface Sci.* 1993, 161,138.

12 G. J. Rodin, *J. Non-Newtonian Fluid Mech.* 1993, 63,141.

13 Y. Xu, W.B.Huang, H.Y.Li, G.P.Lian, *China Particuology* 2005, 3, 52.

14 W.B.Huang, Y.Xu, G.P.Lian,H.Y. Li, *Appl. Math. Mech.* 2002,23722. (in Chinese)

15 G.P.Lian, Y.Xu, W.B.Huang, Adams M J. *J. Non-Newtonian Fluid Mech.* 2001,100,151.

16 H.Y.Li, W.B.Huang,Y. Xu,G.P. Lian, *Particulate Sci Tech.* 2004,22,1.

17 W.B.Huang, H.Y.Li, Y.Xu, G.P.Lian, *Chem Eng Sci.*2006,61,1480.

18 C. H. Xu,W.B Huang,Y. Xu, *Appl. Math. Mech.* 2004, 25,1057. (in Chinese)

PREPARATION AND CHARACTERISTICS OF $LA_xSR_{1-x}COO_3$ AS CATHODE CATALYSTS FOR MICROBIAL FUEL CELL

L.J. Bai, X.Y. Wang, H.B. He, Q.J. Guo

Key Laboratory of Clean Chemical Process, Qingdao University of Science and Technology, Qingdao, 266042, China

1. INTRODUCTION

Microbial fuel cell (MFC) is a kind of renewable technology, which can bring out sewage treatment and generate electricity at the same time.[1-3] However, applications of MFCs have been limited by the lower output voltage and power density. Catalyst for cathode is one of the most key factors influencing the electricity performance of MFCs.[4, 5] For the sluggish oxygen reduction, Pt/C considered good oxygen reduction catalysts, is extremely scarcity and non-renewable.[6, 7] A key challenge to the ultimate commercialization of MFC is the development of steady, active, and cheap catalysts for an oxygen reduction reaction (ORR) in MFC. As a result, new catalysts which could replace precious metal platinum have become the urgent matter.

Oxide materials like precious-metal-free perovskites with excellent oxygen reduction activity are a promising electrocatalyst for cathode in MFC. Perovskite-type oxides with the characteristics of low cost is widely used.[8, 9] Thus, the main purpose of this study was to find out the correlation between electricity generation in MFCs and electrocatalytic activity of the particles. In view of the above, nanometric $La_xSr_{1-x}CoO_3$ (x=0.5、0.7、0.8) perovskite catalysts were synthesized by the sol-gel citrate method. The prepared $La_xSr_{1-x}CoO_3$ catalysts were applied to anaerobic fluidized bed microbial fuel cell (AFBMFC) as the cathode catalysts for investigating its electricity generation performance.

2. METHOD AND RESULTS

2.1 Materials and Methods

$La_xSr_{1-x}CoO_3$ sample was synthesized by sol-gel citrate method. $La(NO_3)_3$, $Sr(NO_3)_2$ and $Co(NO_3)_2 \cdot 6H_2O$ of analytical grade, as the raw materials for metal sources in stoichiometric amount, were crushed. Then these metal nitrates were mixed in deionized water and stirred vigorously at 78 °C. Anhydrous citric acid (99.5%) was then added as the complexing agents. To ensure complete complexation, solution pH was adjusted to 7 by adding $NH_3 \cdot H_2O$ (28%), resulting in a transparent purple aqueous solution. Then the gel was pretreated at 250 °C for 5 h to form a solid precursor and calcined at 800 °C for 4 h. After cooled down, the powder was milled and sieved.

2.2 Electrode Preparation

All the carbon cloths were first cleaned by soaking them in pure acetone. Following treatment, all cloths were washed several times with distilled water before being used in MFCs. Cathodes were coated with one layer of a mixture of acetylene black (2.5 mg/cm^2) and PTFE (60%), and four layers of PTFE (40%) on the air-facing side. Then the catalytic layer was prepared as follows: appropriate amount of $La_xSr_{1-x}CoO_3$ catalysts (loading of 5 mg/cm^2), 200 μL of 5 wt% Nafion and 100 μL isopropanol were made into suspension coated onto the carbon cloth. All electrodes were dried at room temperature for 24 h before used.

2.3 AFBMFC Construction and Operation

The air-cathode single chamber AFBMFC was constructed as described by Kong.[10-12] Briefly, as is shown in Figure 1, the anode chamber (length 60 cm, diameter 4 cm) wired to an external resistor (90000 Ω) was constructed by cylindrical plexiglass with a final volume of 1 L. The cathode was placed on one side of MFC with the oxygen catalyst coating layer facing to the anode directly. The anode were carbon rod (length 350 mm, diameter 7 mm) with a distance of 2.0 cm from the cathode, and no membrane between this two electrodes. The anode chamber partially is filled with coconut shell activated carbon particles, which porosity (ε) is 0.45 and the particle diameter (dp) is from 0.45 - 0.90 mm.

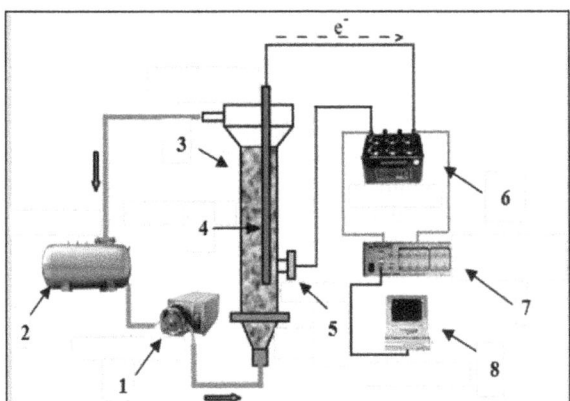

Figure1 *The schematic diagram of anaerobic fluidized bed MFC. 1-peristaltic pump, 2-water storage tank, 3-fluidized anode chamber, 4-anode, 5-air cathode, 6-external resistance, 7-data acquisition system, 8-computer*

Briefly, the working principle of the AFBMFC is described as follows: peristaltic pump transports sewage in reservoir tank to the anode chamber, sewage remain in the anode compartment for a certain period of time, then flow back into the reservoir tank. This is a completely cyclic process.

AFBMFCs were inoculated with domestic wastewater collected from Wastewater Treatment Plant in Qingdao, China. The nutrient buffer solution containing the following (g • L^{-1}): sucrose 1.0, NH$_4$Cl 0.23, CaCl$_2$ 0.123, KCl 0.33, NaCl 0.31, MgCl$_2$ 0.315, K$_2$HPO$_4$ 1.3, KH$_2$PO$_4$ 0.42, yeast extract 1.0. Trace elements was also added into the nutrient buffer solution during the training process.[6] The solution in the anode chamber was replaced when the voltage dropped below 20 mV, forming one complete cycle of operation. The external resistance was fixed at 90000 Ω and all reactors were operated at 30 °C.

2.4 Calculations

The electronical potential across external resistor was recorded every 2 min with a data acquisition system (USB1608 FS, Measurement Computing Corp.). Polarization data were collected by changing the resistance (varied from 20 Ω to 900000 Ω) with a variable resistor box during the stable power production stage of each batch experiment.

The current (mA) of AFBMFC was calculated according to Ohm's law:

$$I = \frac{U}{R} \tag{1}$$

And power density (mW•m^{-2}) was calculated by

$$P = \frac{U^2}{RA} = \frac{UI}{A} \tag{2}$$

where U (mV) is the voltage, P (mW • m^{-2}) is power density, I (mA) is the current, A (cm^2) is the geometric surface area of the cathode electrode, R (Ω) is the external resistance.

2.5 Electrochemical Measurements

Electrochemical measurements were carried on a electrochemical workstation (CS310, Wuhan Corr Test Instrument Co. Ltd.) with a typical three-electrode cell. A glassy carbon (GC) electrode with a diameter of 3 mm was used as the working electrode, a platinum electrode as the counter electrode, and saturated calomel electrode (SCE) as the reference electrode. Prior to use, the bare GC electrode was polished with Al$_2$O$_3$ powders (particle sizes of 0.05 mm).

2.6 Results

To determine the activity of the prepared samples in terms of oxygen reduction, characterization studies of the cathode catalysts were investigated.

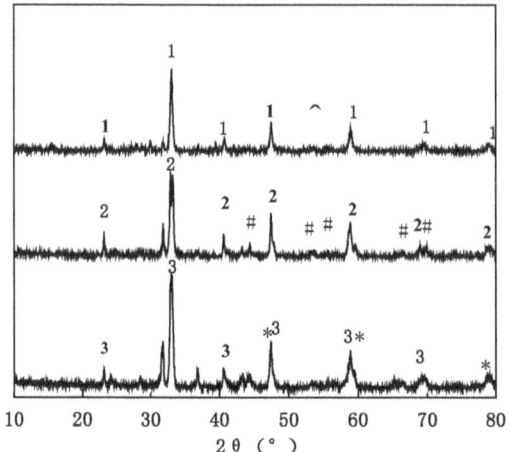

Figure 2 *XRD patterns of different La_xSr_{1-x}CoO₃ samples. 1—La_{0.5}Sr_{0.5}CoO₃,
^—La_{0.6}Sr_{0.4}CoO₃, 2—La_{0.7}Sr_{0.3}CoO₃, #—LaCoO₃, 3—La_{0.8}Sr_{0.2}CoO₃,
—Sr_{0.5}CoLa_{0.5}O_{3-x}.

2.6.1 XRD patterns.
The structures of the nanocrystalline perovskite-type oxides $La_xSr_{1-x}CoO_3$ particles were examined by X-Ray Diffraction (XRD). Figure 2 shows representative XRD patterns for $La_xSr_{1-x}CoO_3$ catalyst.

The results showed that the catalysts made by the sol-gel method have different levels of the perovskite structure on the surface. XRD patterns show very narrow peaks of $La_xSr_{1-x}CoO_3$ (as inferred from the JCPDS data file) without any peaks that can be assigned to its oxides. Note that the typical diffraction peaks of $La_xSr_{1-x}CoO_3$ were observed, which might be ascribed to the crystal structure is controlled at an ideal calcination temperature of 800 °C. Moreover, the average particle size of the prepared $La_xSr_{1-x}CoO_3$ catalysts is in the range of 40-60 nm.

2.6.2 Electrochemical Characteristics.
To investigate the effect of $La_xSr_{1-x}CoO_3$ perovskite catalysts on oxygen reduction reaction (ORR), the electrochemical characteristics were systematically compared by cyclic voltammogram (CV) measurements. As is shown in Figure 3, CV was investigated in phosphate buffered solution (PH=7). All measurements were conducted at 50 mV s^{-1} saturated with O_2 at room temperature.

2.6.3 The Performances of MFCs.
The performances of MFCs with $La_xSr_{1-x}CoO_3$ cathodes were used through gradually decreased external circuit resistances to determine power densities and polarization curves for each of the synthesised catalyst. Figure 5 demonstrated MFC performance in wastewater using a single chamber air cathode MFC.

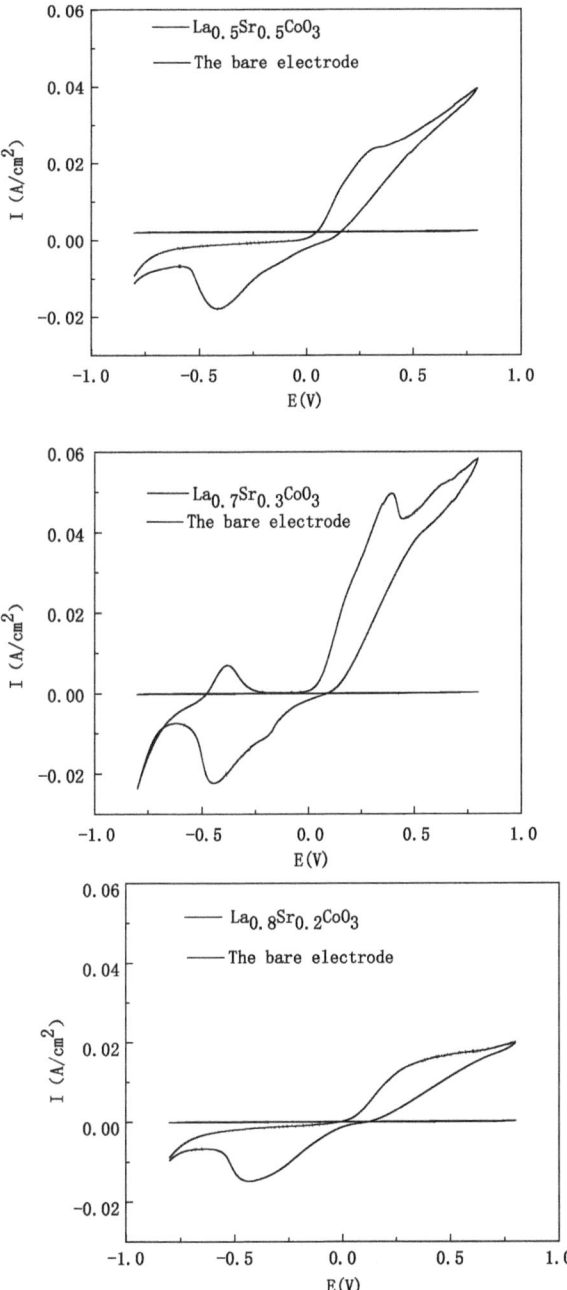

Figure 3 *Cyclic Voltammogram of different air electrodes in phosphate buffered solution*

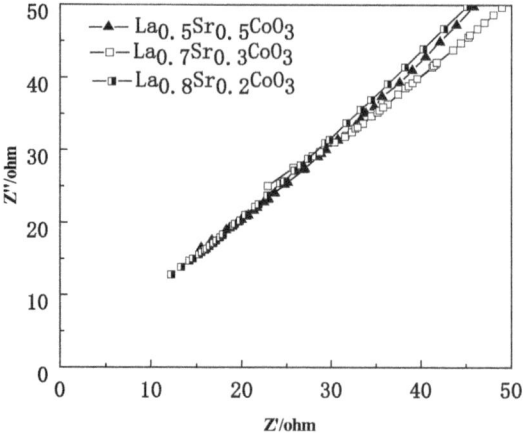

Figure 4 *EIS of different cathode catalysts in phosphate buffer solution (pH=7)*

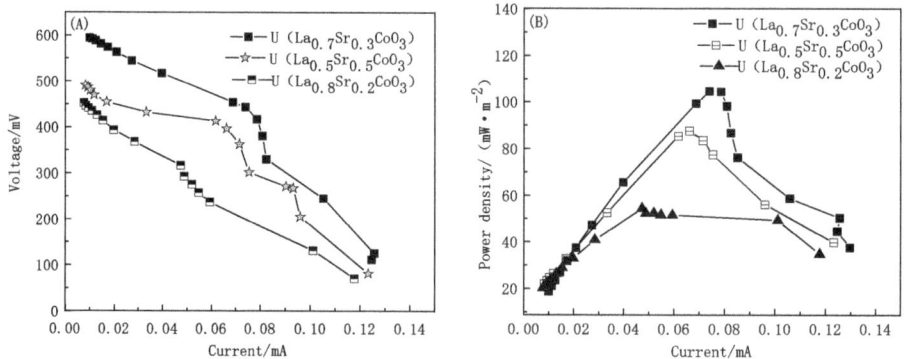

Figure 5 *Polarization curve (A) and power density curve (B) of MFC with different cathode catalysts*

In this study, the cathode was the main focus. The differences from the MFCs were mainly from the cathode side since the same anode was used. The performance of the prepared $La_xSr_{1-x}CoO_3$ catalysts applied for cathode catalysts was examined in the AFBMFCs under the condition of mass transfer reinforcement. The results show that, when the MFCs were operated at steady state, the maximum power of AFBMFC operated with $La_{0.7}Sr_{0.3}CoO_3$ cathode catalysts was 104.59 mW $\cdot$ m^{-2}, which was 20% and 54% higher than AFBMFCs with $La_{0.5}Sr_{0.5}CoO_3$ and $La_{0.8}Sr_{0.2}CoO_3$ catalysts, respectively. The open circuit voltage with $La_{0.7}Sr_{0.3}CoO_3$ catalysts was up to 594.0 mV.

The results demonstrate that the nanometric $La_xSr_{1-x}CoO_3$ perovskite catalysts can lead to even more exotic electrocatalytic activity. These studies showed that a significant activity in oxygen reduction could be achieved with perovskite catalysts. The application of nanometric La$_x$Sr$_{1-x}$CoO$_3$ particles as catalysts instead of expensive Pt in the cathode, probably provides a new solution for finding effective cathode materials for MFC.

3. CONCLUSION

La$_x$Sr$_{1-x}$CoO$_3$ sample was synthesized by sol-gel citrate method and used as cathode catalyst of AFBMFCs. Cyclic Voltammogram showed that improved oxygen reduction activity from La$_{0.7}$Sr$_{0.3}$CoO$_3$ compared to La$_{0.5}$Sr$_{0.5}$CoO$_3$ and La$_{0.8}$Sr$_{0.2}$CoO$_3$ in phosphate buffer solution (PH=7). The application of nanometric La$_{0.7}$Sr$_{0.3}$CoO$_3$ particles as cathode catalysts with power density of 104.59 mW $\cdot$ m^{-2} in wastewater provides a new solution for finding effective cathode materials for MFC.

References

1. B. E. Logan, *Applied microbiology and biotechnology.*, 2010, **85**, 1665.
2. E. H. Yu, R. Burkitt, X. Wang and K. Scott, *Electrochemistry Communications.*, 2012, **21**, 30.
3. S. A. Cheng, B. E. Logan, *Bioresource Technology*, 2011, **102**, 4468.
4. S. Z. Li, Y. Y. Hu, Q. Xu, J Sun, B Hou and Y.P. Zhang, *Journal of Power Sources.*, 2012, **213**, 265.
5. B. Lai, P. Wang, H. R. Li, Z. W. Du, L. J. Wang and S. C. Bi, *Bioresource Technology.*, 2013, **131**, 321.
6. Y.W. Ma, Z.R. Liu, B.L. Wang, L. Zhu, J.P. Yang and X.A Li, *New Carbon Materials.*, 2012, **27**, 250.
7. M. Lefèvre, J. P. Dodelet, *Phys. Chem. B.*, 2000, **104,** 11238.
8. K.E. Johnston, M.R. Mitchell, F. Blanc, P. Lightfoot and S.E. Ashbrook, *Phys. Chem. C.*, 2013, **117**, 2252.
9. K. Ramesha, L. Sebastian, B. Eichhorn and J. Gopalakrishnan, *Chem. Mater.*, 2003, **15,** 668.
10. W. F. Kong, Q. J. Guo, X. Y. Wang and X. H. Yue, *Ind. Eng. Chem. Res.*, 2011, **50**, 12225.
11. S.J. Zhao, X.H. Yue, Q.J. Guo, X.Y. Wang and L.Y. Hou, *Environ. Chem.*, 2010, **29**, 700.
12. C.C. Zheng, Q.J. Guo, X.Y. Wang and W.F. Kong, *CIESC J.*, 2012, **63**, 1599.
13. Y. Ahn, B. E. Logan, *Energy & Fuels.*, 2012, **27**, 271.
14. L. Y. Feng, Y. Y. Yan, Y. G. Chen and L. J. Wang, *Energy Environ. Sci.*, 2011, **4**, 1892.

SIMULATION OF BINARY GROUPS OF PARTICLES WITH A SUB-GRID SCALE MODEL IN A RISER

Wengen Peng, Fenglin Tian, Shenglan Yan and Yurong He

Department of Power Engineering, Harbin Institute of Technology, Harbin 150001, China

1. INTRODUCTION

The gas-solid two phase flow in a circulating fluidized bed (CFB) riser is common and important, mainly because of its wide application in petroleum, chemical and energy industries. It involves highly complex mechanisms both from a strong turbulence of the gas and random collision of the particles.1[1,2] The Discrete Element Method (DEM), originally proposed by Cundall and Strack,[3] has been proved a useful tool for investigating the kinetic characteristic of particle systems. It is widely applied in hopper, fluidized bed, vibrating bed, mixer and hill.[4-12]

An Eulerian and Lagrangian method is used in DEM for the description of the gas phase and solid particles, respectively. The turbulence of the gas is a complex phenomenon.[1] It reflects the velocity fluctuation of the gas and gives rise to viscous diffusion of energy, which finally shows a pronounced effect on the particles' movement. There are three numerical methods to address the flow of the gas phase: Direct Numerical Simulation (DNS), Large Eddy Simulation (LES) and Reynolds Average Navier-Stocks (RANS). The DNS method directly calculates the transient Navier-Stocks equation and consequently it is accurate. Nevertheless, an extremely high resolution is required both in space and time, which takes an unacceptably huge computing time and is only realizable for low-Reynolds-number flow. On the other hand, the RANS method is commonly used in complex flow in engineering. Low computing cost is required but its precision is relatively low and a universality of additional constitute equations for multiphase flow is not achieved yet. The LES method, firstly proposed by Deardoff,[13] becomes a powerful approach for turbulent problems, in which the large-scale turbulent flow is directly predicted by solving the filtered Navier-Stocks equations and the small-scale (sub-grid) fluctuation of the turbulent flow is modeled.[14] It synthesizes the advantages of the DNS method and the RANS method, requiring the lower resolution than the DNS method and supplying much more information of the flow field than the RANS method. Moreover, it is feasible for high Reynolds number flows. Interest of LES method is increasing both in single- and two-phase flow. Ciofalo[15] tested the potentiality of the LES method for the turbulent flow. Yan et al.[16] investigated the effect of two sub-grid scale (SGS) models on a free turbulent round jet flow. Helland et al.[17] studied the group effect of particles and clusters formation in a two-dimensional CFB riser. Yuu et al.[18] numerically studied a group-B particle turbulent fluidized bed using the LES method accounting the existence of particle on SGS flows. Senoner *et al.*[19] carried out LES simulations on evaporation of two-phase flow in a burner. Yin *et al.*[20] proposed two SGS models for the gas viscosity and particle pressure for gas-particles flow in a bubble fluidized bed. Though LES method has already demonstrated its potential for single-phase flow, few among these researchers considered its effect on dynamics of particles in a particle system. More work is needed for its application in two-phase flow.

In CFB riser, particle segregation is common due to different sizes and/or densities and it plays an important role on the flow behavior. In this work, a DEM numerical simulation of a gas-solid flow with two different sizes particles is investigated in a CFB riser. The gas turbulence is considered with the LES method and the anisotropic characteristics of particles are investigated due to more attention paid by researchers now.

2. SIMULATION METHOD

2.1 Gas Phase

The DEM model uses an Eulerian description of the gas. In the LES method, the governing equations are established from filtered Navier-Stockes equations by decomposing the instantaneous gas velocity u_g into a filtered large-scale velocity $\bar{u}_g$ and a small-scale turbulent velocity $u_g^{(s)}$, i.e. $u_g = \bar{u}_g + u_g^{(s)}$. The decomposition procedure is achieved using a low-pass filter,

$$\bar{u}_g(r,t) = \int u_g(\xi,t) G_\Delta(r,\xi) d\xi \tag{1}$$

The small-scale fluctuation in these irregular quantities is filtered out, remaining the turbulent fluctuation larger than the characteristic length Δ, which is given by cell sizes,

$$\Delta = (\Delta x \Delta y \Delta z)^{1/3} \tag{2}$$

The LES governing equations of the mass and momentum are as follows,

$$\frac{\partial}{\partial t}(\varepsilon_g \rho_g) + \frac{\partial(\varepsilon_g \rho_g \bar{u}_{g,i})}{\partial x_i} = 0 \tag{3}$$

$$\frac{\partial}{\partial t}(\varepsilon_g \rho_g \bar{u}_{g,i}) + \frac{\partial}{\partial x_j}(\varepsilon_g \rho_g \bar{u}_{g,i} \bar{u}_{g,j}) = -\varepsilon_g \frac{\partial \bar{p}}{\partial x_i} + \frac{\partial}{\partial x_j}(\varepsilon_g \bar{\tau}_{g,ij}) - F_{p,i} + \varepsilon_g \rho_g g_i \tag{4}$$

where the stress tensor $\bar{\tau}_{g,ij}$ is modeled considering the gas phase as a Newtonian fluid,

$$\bar{\tau}_{g,ij} = (\mu + \mu_t) \cdot \left(\frac{\partial \bar{u}_{g,i}}{\partial x_j} + \frac{\partial \bar{u}_{g,j}}{\partial x_i} - \frac{2}{3} \frac{\partial \bar{u}_{g,k}}{\partial x_k} \delta_{ij} \right) \tag{5}$$

The SGS turbulent viscosity μ_t is calculated with Smagorinsky's model[21] as follows,

$$\mu_t = C_m \Delta^2 \left\langle 2 \overline{S}_{ij} \overline{S}_{ij} \right\rangle^{1/2} \tag{6}$$

where the Smagorinsky constant C_m equal 0.01. The strain tensor $\overline{S}_{ij}$ is given by,

$$\overline{S}_{ij} = \frac{1}{2} \left(\frac{\partial \overline{u}_i}{\partial x_j} + \frac{\partial \overline{u}_j}{\partial x_i} \right) \tag{7}$$

2.2 Solid Particles

In the DEM approach, the motion of an individual particle follows Newton's second law. The translational motion is governed by four forces, i.e. the pressure gradient force, the drag force, the gravitational force and the contact force between colliding particles,

$$m_p \frac{d^2 r}{dt^2} = -V_p \nabla \overline{p} + \frac{V_p \beta}{1 - \varepsilon_g} \left(\overline{u}_g - v_p \right) + m_p g + \left(F_n + F_t \right) \tag{8}$$

Here, the buoyancy force is not considered for high density ratio of particle to gas. The Magnus lift force and Saffman lift force are also neglected due to small particle diameter. A linear spring-dashpot (LSD) model is used in the soft-sphere method to compute the contact forces in the normal and tangential directions between particles and between particles and the wall,

$$F_n = -k_n \delta_n - \eta_n v_n \tag{9}$$

$$F_t = \begin{cases} -k_t \delta_t - \eta_t v_t & |F_t| \leq \mu_f |F_n| \\ -\mu_f |F_n| \frac{v_t}{|v_t|} & |F_t| > \mu_f |F_n| \end{cases} \tag{10}$$

The LSD model is a simple model in which Hooke's law is used for describing the elastic contact force. More complex spring-dashpot models can be found in the works of Schäfer et al.[22] and Walton[23]. The angular velocity under the action of a torque is given as,

$$I_p \frac{d\omega_p}{dt} = T_p \tag{11}$$

2.3 Interphase Exchange

As a function of the product of the interphase momentum exchange coefficient and the relative velocities of the two phases, the rate of exchange of momentum F_p between the particle and the gas phase is computed by the sum of drag forces acting on all individual particles in a computing cell,

$$F_p = \frac{1}{V_{cell}} \sum_{k=1}^{n} \frac{V_p \beta}{1 - \varepsilon_g} \left(\overline{u}_g - v_p^k \right) \tag{12}$$

The interphase momentum exchange coefficient β depends strongly on the local void fraction of the gas phase. According to Beetstra et al.,[24] it can be written as,

$$\beta = K_1\left(\mu+\mu_t\right)\frac{\left(1-\varepsilon_g\right)^2}{d_p^2\varepsilon_g} + K_2\left(\mu+\mu_t\right)\frac{\left(1-\varepsilon_g\right)\mathrm{Re}}{d_p^2} \tag{13}$$

$$K_1 = 180 + 18\frac{\varepsilon_g^4}{1-\varepsilon_g}\left(1+1.5\sqrt{1-\varepsilon_g}\right) \tag{14}$$

$$K_2 = 0.31\frac{\varepsilon_g^{-1}+3\varepsilon_g\left(1-\varepsilon_g\right)+8.4\,\mathrm{Re}^{-0.343}}{1+10^{3\left(1-\varepsilon_g\right)}\,\mathrm{Re}^{2\varepsilon_g-2.5}} \tag{15}$$

where the Reynolds number for the solid phase is defined as,

$$\mathrm{Re} = \frac{\rho_g\varepsilon_g d_p\left|\bar{u}_g-v_p\right|}{\mu+\mu_t} \tag{16}$$

2.4 Particle Velocity and Granular Temperature

At a given time, the cell-averaged mean velocity $\bar{v}_c$ in ith direction may be obtained by,

$$\bar{v}_{c,i}\left(r,t\right) = \frac{1}{n}\sum_{k=1}^n v_{p,i}^k\left(r,t\right) \tag{17}$$

The fluctuating velocity or the Root Mean Square (RMS) velocity v_{RMS} may be expressed as,

$$v_{RMS,i}\left(r,t\right) = \sqrt{\frac{1}{n-1}\sum_{k=1}^n\left[v_{p,i}^k\left(r,t\right)-\bar{v}_{c,i}\left(r,t\right)\right]} \tag{18}$$

The granular temperature, based on the random fluctuation of particle velocity, is one of the most important parameters to measure the kinetic behavior in particulate systems. According to Tartan and Gidaspow[25] and Jung et al.[26,27], the granular temperature, caused by formation and motion of bubbles and giving rise to the normal Reynolds stresses, is related to the second moment of bubble velocity fluctuation,

$$\left\langle V_i V_j\right\rangle\left(r\right) = \frac{1}{m}\sum_{k=1}^m\left[\bar{v}_{c,i}^k\left(r,t\right)-\bar{v}_i\left(r\right)\right]\left[\bar{v}_{c,j}^k\left(r,t\right)-\bar{v}_j\left(r\right)\right] \tag{19}$$

where the time-averaged mean velocity $\bar{v}$ is defined as,

$$\bar{v}_i\left(r\right) = \frac{1}{m}\sum_{k=1}^m\bar{v}_{c,i}^k\left(r,t\right) \tag{20}$$

Finally the time-averaged bubble granular temperature is,

$$\theta(r) = \frac{1}{3}\left[\langle V_x V_x \rangle(r) + \langle V_y V_y \rangle(r) + \langle V_z V_z \rangle(r)\right] \tag{21}$$

2.1 Boundary Conditions

The geometry of the thin CFB riser, shown in Figure 1, is 32 mm wide, 1.2 mm thick and 300 mm high. There are two groups of particles with mean diameters of 120 µm and 185 µm, respectively, in the riser. The total number of particles is 201000, so that the overall volume concentration of particle in the riser is 2.5%. Each group has the same volume concentration. A periodic boundary condition is set for particles. This means no obstacle is set at the entrance or the exit, so that particles are free to leave at the top and re-enter at the bottom of the riser or conversely. The air inlet is modeled as one-dimensional uniform flow and three different superficial gas velocities are considered, i.e. 0.8 m/s, 1.0m/s and 1.2 m/s, respectively. A total pressure of 1.2×10^5 Pa is set at the outlet. For the gas phase, a full-slip boundary condition is applied for the two walls in the thickness direction and a no-slip boundary condition is defined for the two walls in the width direction.

A modified MFIX-DEM code (Multiphase Flow with Interphase eXchanges),[28] developed at NETL (National Energy Technology Laboratory), is used for simulations in the present work. The finite volume approach is applied with discretization on a staggered grid for the hydrodynamic model. The well-known SIMPLE scheme is used as iterative solution procedure.[29] All simulations are carried out for 10 s and the time-averaged results are obtained from the last 4 s. The simulation results are compared to those measured data from Mathiesen.[30] The structure sizes and properties of gas and particles used in the simulation are summarized in Table 1.

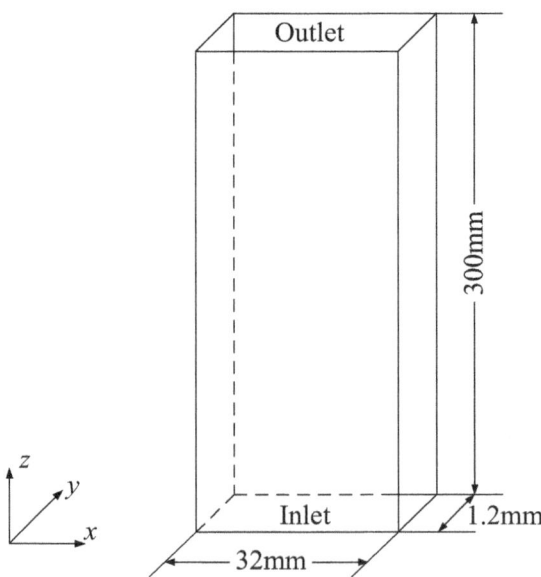

Figure 1 *Structure scheme of the CFB riser*

Table 1 *Parameters applied in the simulation*

VARIABLE	VALUE	UNIT	VARIABLE	VALUE	UNIT
Bed			*Particle*		
Width, x	32	mm	Number of particles	201000	-
Thickness, y	1.2	mm	Particle diameter, d_p	120,185	μm
Height, z	300	mm	Particle density	2400	kg/m^3
Cells in x, y, and z direction	64×3×375	-	Coefficient of restitution, e	0.97	-
Superficial gas velocity, U	0.8, 1.0, 1.2	m/s	Coefficient of sliding friction, μ_f	0.10	-
Outlet pressure	1.2×10^5	Pa	Normal spring stiffness, k_n	100	N/m
			Tangential spring stiffness, k_t	28.6	N/m
Gas					
Temperature	298	K			
Density, ρ_g	1.2	kg/m^3			
Viscosity, μ	1.8×10^{-5}	Pa·s			

3. RESULTS AND DISCUSSION

3.1 Particle Flow Structure

Two instantaneous flow structures of gas and particles in the riser are illustrated in Figure 2 at 8.0 s and 9.0 s. The four figures, in each group, are the spatial distribution of particles (Red for big particles, Blue for small particles), the particle volume concentration, the gas velocity and the particle velocity, respectively. It can be seen that particles are distributed heterogeneously and clusters are formed at various levers of the riser, especially at the bottom and near the side walls. With the time goes, the clusters grow up, change their shapes and finally break up. The gas flows upward mainly through the regions of high porosity at the center and the particles, driven by the gas flow, have a similar velocity distribution to that of the gas. Near the walls, the volume concentration of the particles is much higher. The particles mainly settle down in the form of cluster and back flow of the gas exists under the influence of these settling particles.

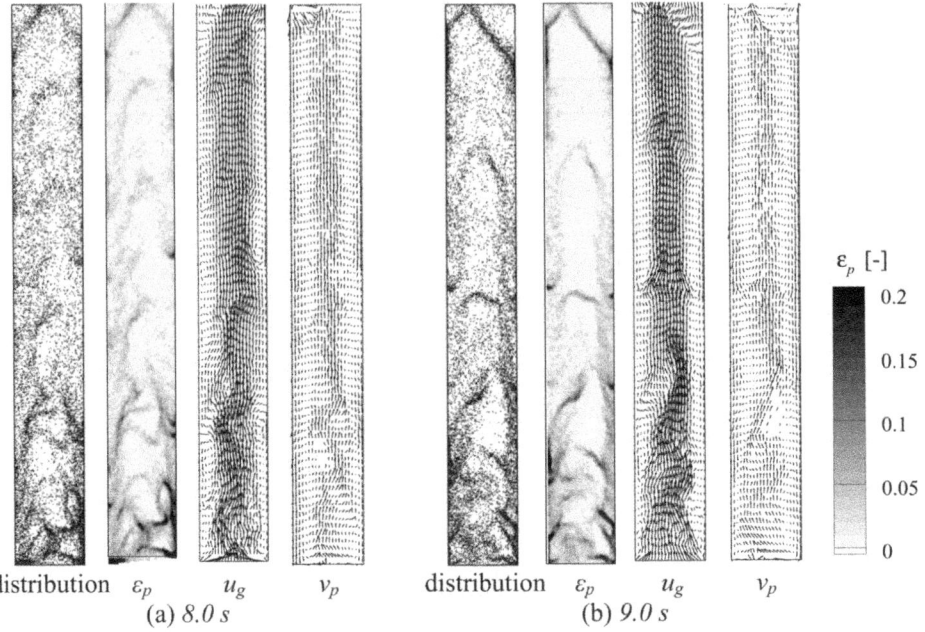

distribution ε_p u_g v_p distribution ε_p u_g v_p

(a) *8.0 s*   (b) *9.0 s*

Figure 2 *Instantaneous gas and particle flow structures*

In Figure 3, the time-averaged velocity profile of each particle group is presented. In the horizontal direction, the particles at the center move up driven by the gas, while the particles near the wall fall down with a negative velocity. This is consistent with a direct observation of particle movement in the riser as shown in Figure 2. Due to obstruction of clusters and high volume concentration of particles, the velocity of the gas near the walls is either low, not enough to sopport the particles, or even negative, which makes the downward velocity be high for particles. There is a separation between the binary groups of particles. The small particles are lighter and easier to follow the flow of gas with a faster movement. This phenomenon is more obvious in the center, where the gas has the highest velocity. Unfortunately, with an increase of the height in the riser, the velocity of particles are overestimated by the present simulation to that of Mathiesen's experiment.[30] This may be because our simulation structure is not exact what the experiment is. In the experiment, the riser is 1.0 m long, while in our simulation it is 0.3 m and the same dimensionless height is considered.

Figure 4 shows the mean particle volume concentration profile at three dimensionless heights (h/z=0.2, 0.4 and 0.7) under a superficial velocity of 1.0 m/s. Near the distributor, the concentration of particles is much higher than the other two heights in the riser. The structure of clusters formed here is more intense and larger. The simulation of the present work and that of Mathiesen[30] both predicted the same trend of heterogeneous profile in the horizontal direction with the experiment of Mathiesen[30], in which the volume concentration is low at the center and high near the walls. It indicates that the particles tend to move from the center to the side walls. Moreover, the simulation in this work gives a more satisfying result to the experiment especially at higher lever in the riser.

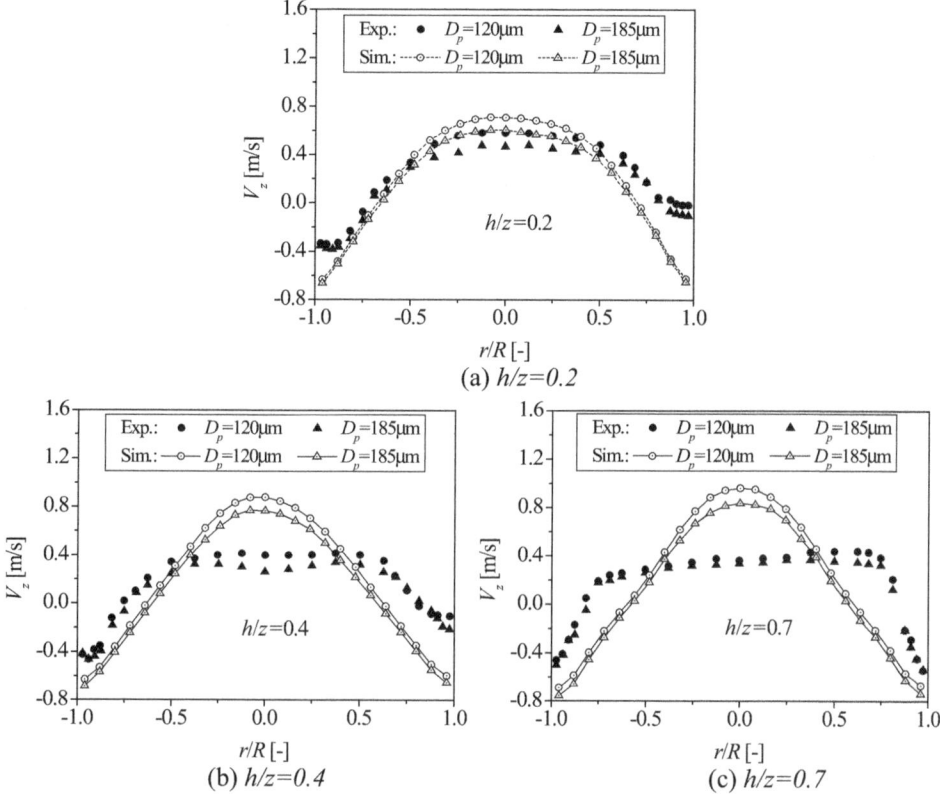

(a) *h/z=0.2*

(b) *h/z=0.4*

(c) *h/z=0.7*

Figure 3 *Velocity profile of each particle group at different heights (U=1.0 m/s)*

Figure 5 shows the time-averaged diameter profile of the whole particle group at three dimensionless heights. The simulation result of the present work is compared to experiment and simulation of Mathiesen[30]. The mean particle diameter in the center is much lower than that near walls. From the figure, it shows that the simulation of Mathiesen gives an unreasonable profile (low near the wall and high in the center), which is completely opposite to the measure data. Although the simulation of the present work underestimated the diameter near walls, it predicted a good trend with the measure data, i.e. a increase of mean diameter from the center to the walls. The mean particle diameter decreases slightly with the increase of the height. This means small particles are easily carried away by the gas, leaving the big particles accumulating at a lower height.

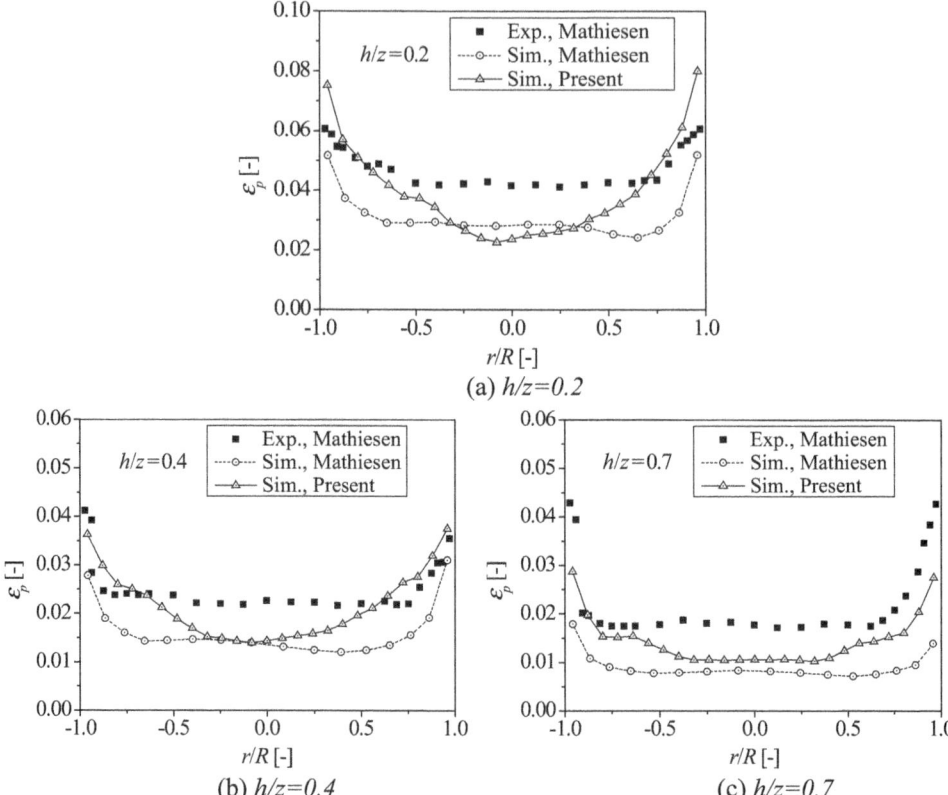

(a) *h/z=0.2*

(b) *h/z=0.4*   (c) *h/z=0.7*

Figure 4 *Transverse profile of mean particle volume concentration (U=1.0 m/s)*

A comparison of the RMS velocity is conducted between the simulation in this work and the Mathiesen's experiment for each particle group in Figure 6. The RMS velocity increases to a maximum value and then decreases rapidly from the center toward the side walls and decreases slightly with the increase of the height. The turbulence of small particles are more dense than big particles at *h/z*=0.2, where much more particles are accumulated in this dense region and the particle/particle collision has a strong influence on this. The difference between the turbulence of small particles and that of big particles decreases at higher lever in the riser. The simulation of the present work failed to predict this phenomenon.

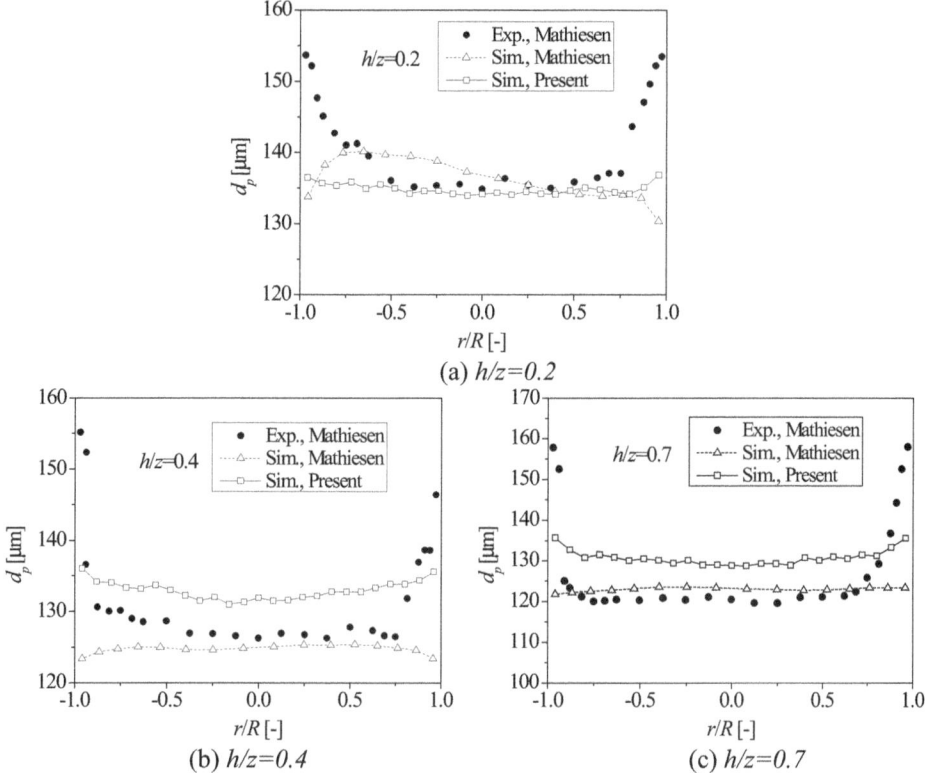

Figure 5 *Transversal profile of mean particle diameter (U=1.0 m/s)*

The ratio of granular temperature θ_{zz} to θ_{xx} is shown in Figure 7 at three heights. As we can see, a strong anisotropic behavior of particle velocity fluctuating is observed in different directions. In the center, the ratio maintains a relative constant value and the granular temperature in the vertical direction is near one order of magnitude greater than that in the horizontal direction. In the near wall regions, this ratio increase rapidly to a astonish value. This may be because the wall restricts the movement of particles in the transvers direction and makes the velocity of particles near zero. From the figure, it indicates that the LES with the DEM model is able to predict the anisotropic characteristic of particles.

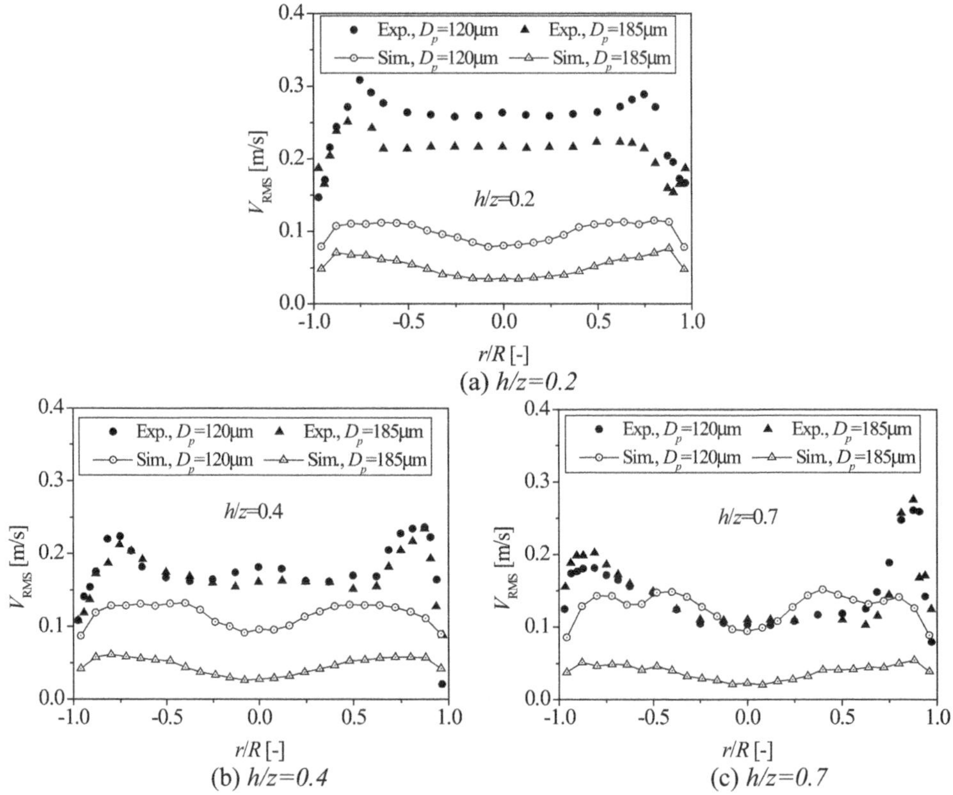

Figure 6 *Root Mean Square velocity of particles (U=1.0 m/s)*

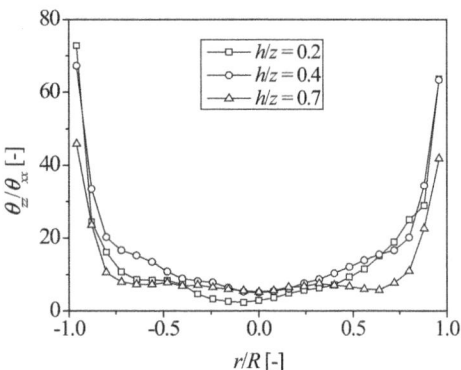

Figure 7 *Transversal profile of ratio of granular temperature in different directions (U=1.0 m/s)*

Figure 8 shows the mean granular temperature as a function of particle volume concentration with a superficial gas velocity of 1.0 m/s. The mean granular temperature first increases and then decrease with the increase of particle volume concentration. A

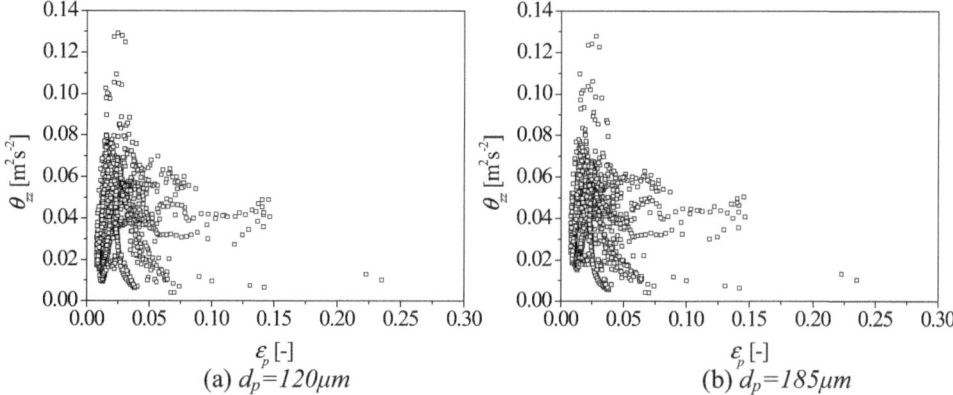

Figure 8 *Granular temperature as a function of particle concentration (U=1.0 m/s)*

maximum value of 0.14 is reached at a particle concentration of 0.025 for both particle groups. Most particles in the riser range from 0 to 0.07 in volume fraction and from 0-0.08 in granular temperature. Difference between the small particles and the big particles is little.

3.2 Effect of Inlet Gas Velocity

Figure 9 shows the velocity profiles at the dimensionless height of 0.4 under the superficial gas veloctiy of 0.8 m/s and 1.2 m/s. Together with Figure 3 (b), the trend of particle velocity is the same and a typical "core-annulus" flow exists in the riser, in which particles move up at the center and fall down near the walls. From the figure, it indicates that the superficial gas velocity has a pronounced effect on the magnitude of particle velocity of both particle groups but a nominal effect on the difference between them.

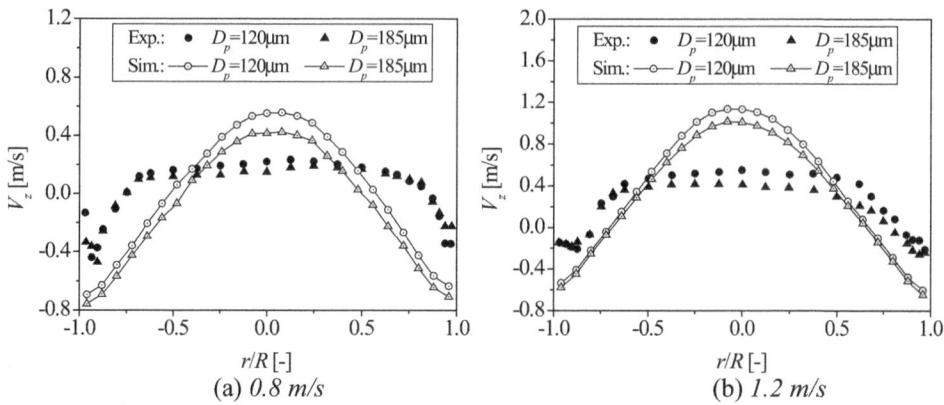

Figure 9 *Velocity profile under different superficial gas velocities (h/z=0.4)*

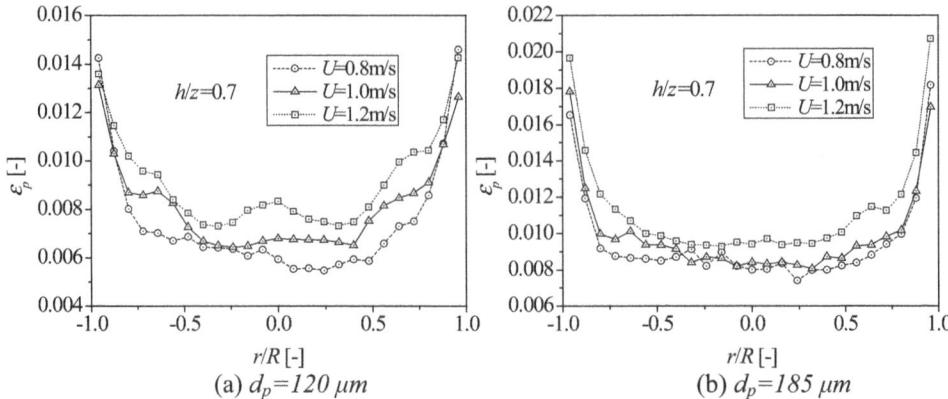

Figure 10 *Particle volume concentration under different superficial gas velocities*

Figure 10 shows the simulated particle volume concentration with superficial gas velocity of 0.8 m/s, 1.0 m/s and 1.2 m/s for each particle group at the dimensionless height 0.7. The volume fraction of small particles and big particles both increase with the increase of the superficial velocity. This means a stronger carrying capacity of the gas brings more particles of both groups from the bottom toward the top of the riser. Particles distributed more uniformly in the riser.

Figure 11 and Figure 12 show the mean granular temperature under three different superficial gas velocities. In Figure 11, the trend of the ratio of the mean granular temperature is still the same and it increases slightly with the increase of the superficial gas velocity at different heights. In Figure 12, it shows that a higher inlet velocity gives a litter more intense maximum granular temperature. The particle concentration region of most particles compresses with a more uniform distribution of particles in the riser. The superficial gas velocity also influence the region of the granular temperature distribution, i.e. the granular temperature with a particle concentration ranging from 0 to 0.23 at 1.2 m/s, while it ranging from 0 to 0.16 at 0.8 m/s. Compared with the granular temperature of the binary groups of particles, the difference is little.

4. CONCLUSION

A LES simulation compiled with the DEM model was carried out to investigate the flow behavior of a binary group of particles with different diameters in a CFB riser. The gas turbulence is considered with the Smagorinsky's SGS model. Clusters were formed mainly at the bottom and near the walls in the riser. Separation between different groups of particles was observed especially in the center. A high ratio of vertical granular temperature to that in horizontal direction is found in the riser, which means a strong anisotropy behavior of particles. A higher superficial gas velocity slightly increases the particle velocity and volume concentration at different height. The region of volume concentration compresses and particles are more uniformly distributed.

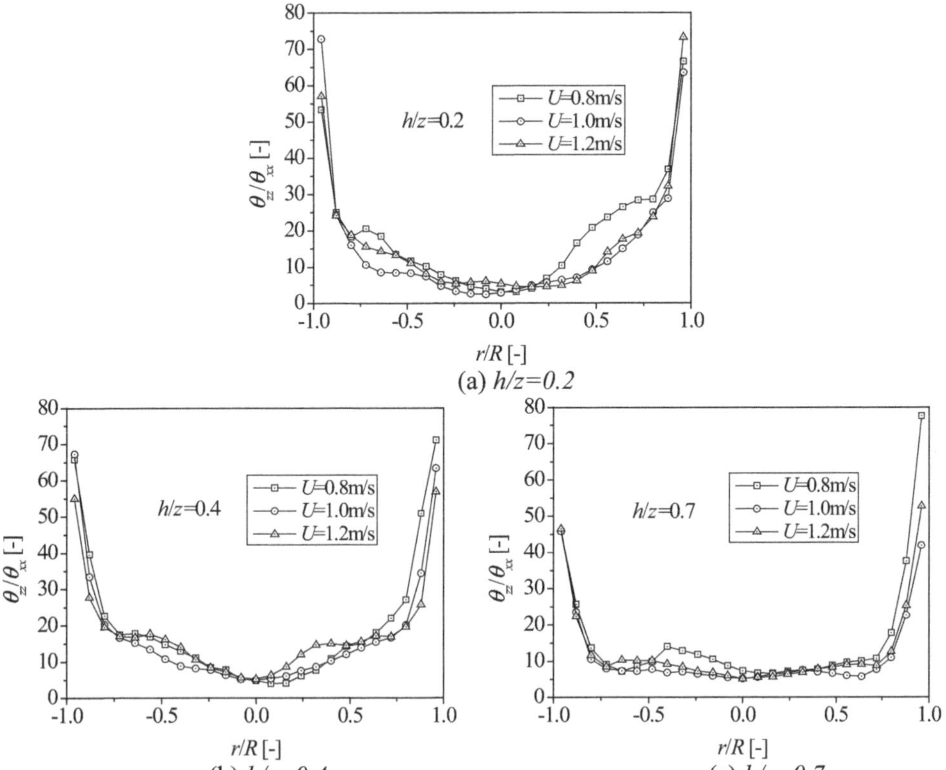

(a) *h/z=0.2*

(b) *h/z=0.4*

(c) *h/z=0.7*

Figure 11 *Ratio of granular temperature under different superficial gas velocities*

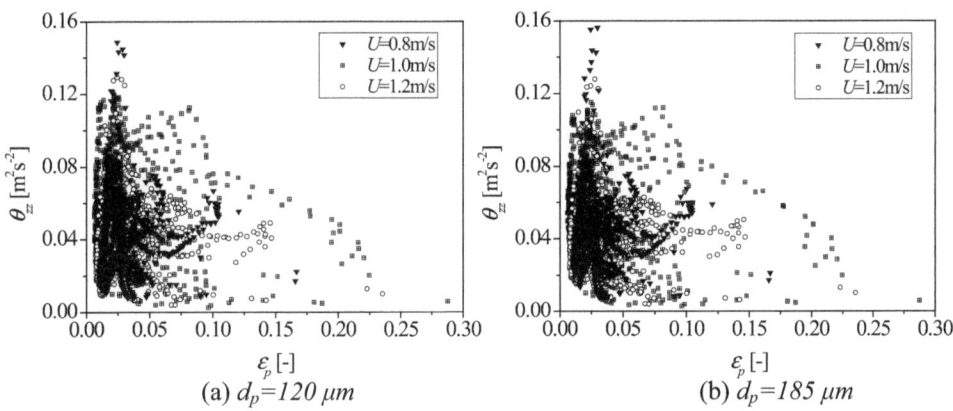

(a) *dₚ=120 μm*

(b) *dₚ=185 μm*

Figure 12 *Granular temperature under different superficial gas velocities*

ACKNOWLEDGEMENTS

This work is financially supported by the National Natural Science Foundation of China-China National Petroleum Corporation Joint Fund of Petrochemical Engineering (Grant No. U1162122).

NOMENCLATURE

C_m	Smagorinsky constant		v	velocity of a particle, $m \cdot s^{-1}$
d_p	particle diameter, m		v_{RMS}	root mean square velocity, $m \cdot s^{-1}$
e	coefficient of restitution			
F_p	rate of exchange of momentum, N		V_{cell}	volume of a computing cell, m^3
$\boldsymbol{F}$	collision force, N		V_p	volume of a particle, m^3
g	gravity, $m \cdot s^{-2}$			
I_p	moment of inertia, $kg \cdot m^2$		**Greek**	
k	spring stiffness, $N \cdot m^{-1}$		β	interphase momentum exchange coefficient
m	number of frames in a time serial		δ	overlap, m
m_p	mass of a particle, kg		δ_{ij}	Kronecker delta
n	number of particles in a cell		Δ	characteristic length, m
$\bar{p}$	gas pressure, Pa		$\Delta x, \Delta y, \Delta z$	cell sizes at different directions, m
$\boldsymbol{r}$	position of the gas or particle center, m		ε_g	void fraction of the gas
Re	Reynolds number		ε_p	particle volume concentration
$\bar{S}_{ij}$	strain tensor, s^{-1}		η	damping coefficient, $N \cdot s \cdot m^{-1}$
t	time, s		θ	granular temperature, $m^2 \cdot s^{-2}$
$\boldsymbol{T}_p$	torque, $N \cdot m$		μ	shear viscosity, $Pa \cdot s$
$\boldsymbol{u}_g$	instantaneous velocity, $m \cdot s^{-1}$		μ_f	coefficient of sliding friction
$\bar{\boldsymbol{u}}_g$	large-eddy filtered velocity, $m \cdot s^{-1}$		μ_t	sub-grid turbulent viscosity, $Pa \cdot s$
$\boldsymbol{u}_g^{(s)}$	small-scale turbulent velocity, $m \cdot s^{-1}$		ρ_g	gas density, $kg \cdot m^{-3}$
			$\bar{\tau}_g$	stress tensor, Pa
U	superficial gas velocity, $m \cdot s^{-1}$		ω_p	rotational velocity, s^{-1}
$\bar{v}$	time-averaged mean velocity, $m \cdot s^{-1}$		**Subscript**	
			n	normal
$\bar{v}_c$	cell-averaged mean velocity, $m \cdot s^{-1}$		t	tangential

References

1. E. Peirano and B. Leckner, *Prog. Energ. Combust.*, 1998, **24**, 259.
2. H. Zhou, J. Lu and L. Lin, *Chem. Eng. Sci.*, 2000, **55**, 839.
3. P.A. Cundall and O.D.L. Strack, *Geotechnique*, 1979, **29**, 47.

4. R. Balevičius, R. Kačianauskas, Z. Mróz and I. Sielamowicz, *Adv. Powder Technol.*, 2011, **22**, 226.
5. Y. Tsuji, T. Kawaguchi and T. Tanaka, *Powder Technol.*, 1993, **77**, 79.
6. B.P.B. Hoomans, J.A.M. Kuipers, W.J. Briels and W.P.M. Van Swaaij, *Chem. Eng. Sci.*, 1996, **51**, 99.
7. B.H. Xu and A.B. Yu, *Chem. Eng. Sci.*, 1997, **52**, 2785.
8. F. Alobaid, J. Ströhle and B. Epple, *Adv. Powder Technol.*, 2013, **24**, 403.
9. S.S. Hsiau and S.C. Yang, *Chem. Eng. Sci.*, 2003, **58**, 339.
10. H.P. Kuo, P.C. Knight, D.J. Parker, M.J. Adams and J.P.K. Seville, *Adv. Powder Technol.*, 2004, **15**, 297.
11. J.J. McCarthy, D.V. Khakhar and J.M. Ottino, *Powder Technol.*, 2000, **109**, 72.
12. T. Iwasaki, T. Yabuuchi, H. Nakagawa and S. Watano, *Adv. Powder Technol.*, 2010, **21**, 623.
13. J.W. Deardorff, *J. Fluid. Eng.*, 1973, **95**, 429.
14. M. Lesieur and O. Metais, *Annu. Rev. Fluid Mech.*, 1996, **28**, 45.
15. M. Ciofalo, *Appl. Math. Model.*, 1996, **20**, 262.
16. H. Yan and M. Su, *Commun. Nonlinear Sci.*, 1999, **4**, 12.
17. E. Helland, R. Occelli and L. Tadrist, *Powder Technol.*, 2000, **110**, 210.
18. S. Yuu, H. Nishikawa and T. Umekage, *Powder Technol.*, 2001, **118**, 32.
19. J.M. Senoner, M. Sanjosé, T. Lederlin, F. Jaegle, M. García, E. Riber, B. Cuenot, L. Gicquel, H. Pitsch and T. Poinsot, *Comptes Rendus Mécanique*, 2009, **337**, 458.
20. L.J. Yin, S.Y. Wang, H.L. Lu, S. Wang, P.F. Xu, L.X. Wei and Y.R. He, *Chem. Eng. Sci.*, 2010, **65**, 2664.
21. J. Smagorinsky, *Mon. Weather Rev.*, 1963, **91**, 99.
22. J. Schäfer, S. Dippel and D.E. Wolf, *J. phys. I*, 1996, **6**, 5.
23. O.R. Walton, *Particulate Two-phase Flow*, ed. M.C. Roco, Butterworth-Heinemann, London, 1992, ch. 25, p. 884.
24. R. Beetstra, M.A. Van der Hoef and J.A.M. Kuipers, *AIChE J.*, 2007, **53**, 489.
25. M. Tartan and D. Gidaspow, *AIChE J.*, 2004, **50**, 1760.
26. J. Jung, D. Gidaspow and I.K. Gamwo, *Ind. Eng. Chem. Res.*, 2005, **44**, 1329.
27. J. Jung, D. Gidaspow and I.K. Gamwo, *Chem. Eng. Commun.*, 2006, **193**, 946.
28. M. Syamlal, W. Rogers and T.J. O'Brien, *NETL Tech. Rep.*, 1993, DOE/METC-94 /1004.
29. S.V. Patankar, *Numerical Heat Transfer and Fluid Flow*, Taylor & Francis, New York, 1980.
30. V. Mathiesen, T. Solberg and B.H. Hjertager, *Int. J. Multiphas. Flow*, 2000, **26**, 387.

VALIDATION OF POWDER PROPERTIES MEASURED BY A ROTATIONAL SHEAR CELL

T. Freeman and X. Fu

Freeman Technology Ltd., 1 Miller Court, Severn Drive, Tewkesbury, UK
E-mail:xiaowei.fu@freemantech.co.uk

1 INTRODUCTION

Powders will be subjected to consolidation stress during their handling, whether it is transport in a keg, storage in a hopper or processing through an IBC on the top of a tablet press or roller-compactor. To understand the relationship between consolidation stress and the ability of a powder to transition from static to incipient flow conditions, a shear cell is commonly used.

The use of rotational shear cells has become commonplace compared to the translational Jenike shear cell[1], due to the advantages of easy operation, quicker testing time and unlimited shear displacement. In a rotational shear cell, a vaned head is used to induce a precisely controlled normal stress on the powder bed. While maintaining this normal consolidating stress, the head then rotates to induce a rotational stress in a shear zone created just below the vanes[2,3]. As the powder initially resists this movement, the shear stress increases until this resistance is overcome and the powder bed fails or shears. This is the point where flow occurs and is known as the Point of Incipient Failure (shear point). The shear stress and the normal stress at the shear points are recorded and the cycle is repeated at a series of normal stress levels.

The measured stress values are used to plot a Yield Locus from which several parameters can be derived by fitting the appropriate Mohr stress circles. Of these parameters, the one most commonly used to describe the 'flowability' of the test sample is Flow Function (FF) which considers the Unconfined Yield Strength (UYS or σ_c) of a powder with respect to the Major Principal Stress (MPS or σ_1) under which the powder is consolidated.

FF values have traditionally been used as key indicators of powder flowability during hopper flow analysis. However, as a parameter derived from the Yield Locus following analysis of shear test data, FF is therefore significantly influenced by how the Yield Locus is drawn. The impact of Yield Loci fitting methods on the derived FF has been investigated by other researchers[4], however, for rotational shear testers, which fitting method generates FF values that represent true powder flow characteristics is still not clear, without comparing to those obtained by direct measurement.

In this study, data collected from a rotational shear cell has been validated by comparing to the certified shear stress values for the standard reference material, BCR 116 limestone[5]. To further support this investigation, a uni-axial testing method, based on the protocol derived by Williams[3,5-7], has been employed to compare directly measured UYS values with those derived from a rotational shear cell as well as to the standard values. The influence of the fitting algorithms used to derive the FF values has then been investigated.

2 MATERIALS AND METHODS

The standard shear cell testing is performed by using the FT4 Powder Rheometer® shear cell accessory (Freeman Technology, Gloucestershire, UK). The FT4 Powder Rheometer® is a universal powder tester, which measures dynamic flow, shear and wall friction properties as well as bulk behaviour[2].

Figure 1   *FT4 Powder Rheometer and its shear cell accessory*

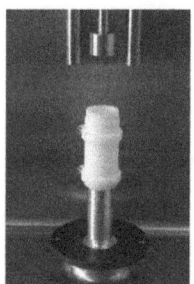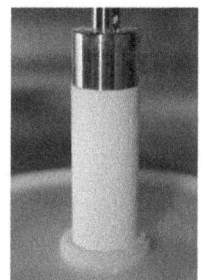

Figure 2   *The uniaxial accessory for FT4 Powder Rheometer (left), BCR116 powder column formed by incremental pre-consolidation (middle) and catastrophic failure pattern with H/D ratios =2.4 (right),*

Uni-axial testing is undertaken with an accessory for the FT4[7] which consists of a stainless sample holder and a Delrin® cylindrical split mould, which slides over the sample holder freely only when two clamp rings are loosened, as shown in Figure 2. This sliding mould effectively reduces disturbance to produce a free-standing powder column. Wall friction effects are overcome by consolidating the powder sample layer by layer using Williams' protocol[6]. The ratio of height and diameter (H/D) of the powder column is higher than tan $(45°+\varphi/2)$ to minimise the influence of the H/D on the measured UYS values. This configuration and experimental protocol produce UYS values comparable with the certified values for the standard reference material generated by the Jenike translational shear tester[3,6-8].

Two different materials are investigated in this study, including BCR116 Limestone (Commission of the European Communities, 4μm, angular) and commercial Talc powder (20μm, platelets).

3 RESULTS AND DISCUSSIONS

3.1 Validation of shear stressed by the rotational shear cell

The only information in the public domain relating to the accuracy and acceptability of shear cell devices is found in Akers[5]. Shear Stresses at the incipient failure points collected from the FT4 rotational shear cell have shown very good agreement with the certified shear stress values for the standard reference material under different, standard pre-consolidation levels (3, 6, 9 and 15kPa respectively), as shown in Figure 3. This confirms the accuracy of the Shear Stress values directly measured by the rotational shear cell employed in this study.

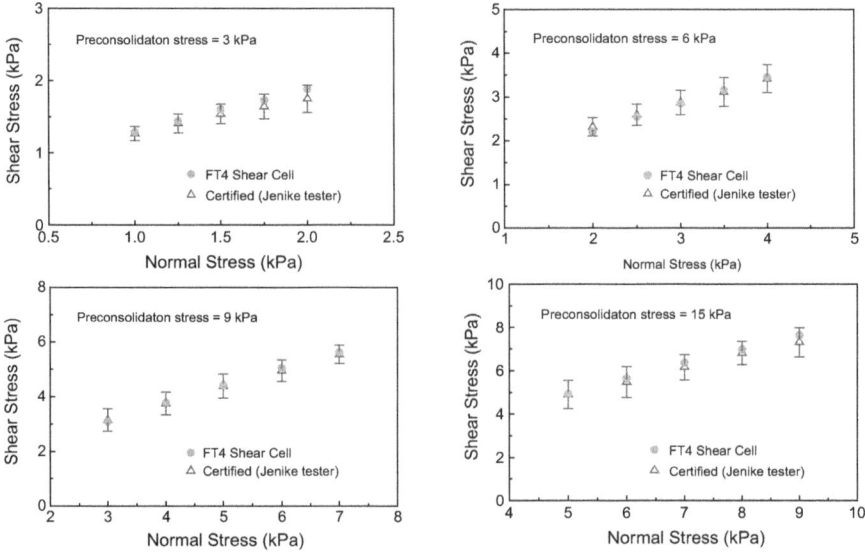

Figure 3   *Comparing the yield loci measured by FT4 shear cell and certified values for BCR116 powder under different pre-consolidation stresses.*

3.2 Data derived from shear testing

Analysis by hand and the construction of a 'smooth' line through the shear stresses at incipient failure points is outdated, either for translational shear cell or rotational shear cell, and most users employ computer based methods to analyse shear data[2,3]. The Yield Locus constructed from shear stresses is commonly a concave curve towards the normal stress axis, particularly for cohesive materials[5]. Drawing the Yield Locus by best linear fit or curved fit can introduce variation in the derived results, as illustrated in Figure 4.

As the uni-axial tester has been proved to generate UYS values comparable with the certified values obtained by the Jenike translational shear tester for BCR116 powder[6,7], this tester can therefore be employed as a validation tool for the evaluation of UYS values generated by different fitting algorithms for rotational shear cells.

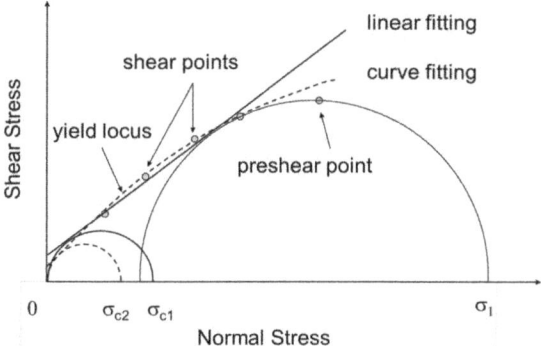

Figure 4 *Shear test data derived from rotational shear testing*

3.3 Linear fitting vs. curved fitting

Mathematically both linear fitting and quadratic fitting are suitable to draw the Yield Locus through the shear stress points for both samples used in this study, with the coefficient of determination being very close to unity, as shown in Table 1.

Table 1. R^2 *values from different Yield Loci fitting algorithms for BCR116 and talc powder over a range of consolidation stresses*

R^2	3kPa	6kPa	9kPa	12kpa	20kpa	30kPa
BCR116, linear	0.9995	0.9995	0.9995	0.9983	0.9984	0.9975
BCR116, quadratic	0.9999	1.0000	0.9998	1.0000	0.9994	0.9999
Talc, linear	0.9999	0.9991	0.9985	0.9977	0.9979	0.9983
Talc, quadratic	0.9999	0.9999	1.0000	0.9999	1.0000	0.9999

However, the derived values from the shear testing methodology still vary significantly with the fitting algorithm used, with the UYS values derived from polynomial fitting in particular being significantly smaller than those from linear fitting.

Figure 5a shows that the UYS values of BCR116 powder derived from a linear fitting algorithm are in line with those directly measured by the uni-axial tester, as well as the certified values, which also use a linear fitting algorithm.

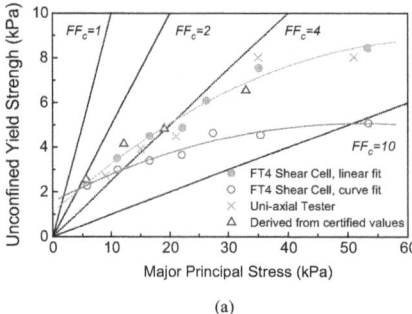
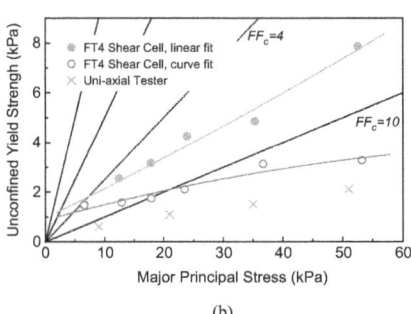

| (a) | (b) |

Figure 5   *Comparing the flow functions for BCR116 (a) and talc (b) samples measured by uniaxial tester and FT4 shear cell with different fitting algorithms.*

It is therefore reasonable to believe that linear fitting is more suitable for cohesive powders. For powders as cohesive as BCR116 limestone, using the linear fitting in the Mohr Circle analysis also has an advantage in hopper design, as it generates lower FF values which lead to larger (safer) outlet sizes.

Conversely, for the talc sample, quadratic fitting for the Yield Loci generates UYS values closer to those measured by uni-axial testing. However, there is still a significant difference between the UYS values generated by the two techniques and further work is required to understand the reason behind this phenomenon.

4 CONCLUSIONS

Parameters derived from shear testing, such as UYS and FF, are significantly influenced by the use of linear and quadratic fitting algorithms, despite the fact that both algorithms generate a coefficient of determination close to unity ($R^2 \approx 1$).

For cohesive powders such as BCR116 limestone, UYS values determined by using a linear fit during Mohr Circle analysis compare to those directly measured by a uni-axial tester, and both are significantly higher than UYS values derived by a quadratic (curve fit) function.

For easy flowing or free flowing powders such as talc powder, the linear fitting algorithm generates substantially higher UYS values than those measured by the uni-axial tester, particularly at a higher stress range.

The evaluation of other materials is ongoing however, the results in this study clearly indicate that different Yield Loci fitting algorithms must be considered for materials with different levels of cohesivity when analysing shear data. Furthermore, it is also clear that the Flow Function parameter has limitations when used solely to categorise powder flowability in general without simultaneously presenting the underlying data.

References

1   A.W. Jenike, Gravity flow of bulk solids, Utah University Engineering Experiment Station Bullletin 123, 1964.
2   R. Freeman, Measuring the flow properties of consolidated, conditioned and aerated powders - a comparative study using a Powder Rheometer and a rotational Shear Cell, Powder Technology, 2007, **174**, 25.
3   D. Schulze, Powder and bulk solids, Springer, Berlin (D), 2008.
4   R.J. Berry and M.S.A Bradley, Investigation of the effect of test procedure factors on the failure loci and derived failure functions obtained from annual shear cells, Powder Technology, 2007, **174**, 60.
5   R.J. Akers, The certification of limestone powder for Jenike shear testing, CRM116. Commission of the European Communities (ECSC-EEC-EAEC), Luxembourg, 1992.
6   J.C. Williams, A.H. Birkes and D. Bhattacharya, The direct measurement of the failure function of a cohesive powder, Powder Technology, 1971, **4**, 328.
7   T. Freeman, X. Fu, The development of a compact uniaxial tester, Particulate Systems Analysis, Edinburgh, 2011.
8   T.A. Bell, Evaluation of Edinburgh Powder Tester, PARTEC, Nuremburg, Germany, 2007.

THREE-DIMENSIONAL SIMULATION OF THE FILTRATION PROCESS OF POLYDISPERSE PARTICULATE MATTER BY FIBROUS FILTER

Kun Wang, Haoming Wang, Haibo Zhao[*], Chuguang Zheng

State key Laboratory of Coal Combustion, Huazhong University of Science and Technology, China
[*]klinsmannzhb@163.com

1. INTRODUCTION

Fibrous filtration, which is at advantage of high capture efficiency of submicron particle, is very widely used in coal-fired power plants, mining engineering, cement industries and indoor air purification. The important collection mechanisms of solid particles are Brownian diffusion, interception, and inertial impaction, as well other mechanisms due to other external forces such as electrostatic forces and gravity. During the past fifty years, many researchers have investigated the filtration process through theoretical analysis[1-4], numerical simulation[5-7] and experimental measurement[8, 9]. A series of idealized models and (semi-)empirical expressions were presented to calculate pressure drop and capture efficiency dominated by various capture mechanisms[1-4]. Generally speaking, these available models/expressions have been validated to be able to characterize the steady-state filtration process of single cylindrical fiber (or a row of regularly arranged fibers) well[1, 7, 10].

Filter cloth is often used to remove the suspended and fine particles. It consists of many fibers through different weaving methods (plain weave, twill weave and satin weave, see Figure 1). Various weaving patterns are commonly characterized by the same micro-structure made by two orthogonal cylinder fibers (see Figure 1). So far, there are a few studies on the woven filter cloth. Lantermann and Hänel [11] used particle Monte Carlo method to simulate the structure of deposited particles and Lattice-Boltzmann method to describe the flow field, and they studied the particles (1~80nm) capture process due to two cylinder fibers whose axes are orthogonal and coplanar. However, they only considered the monodisperse particles in their simulation. Actually, the suspended particles are always polydisperse and their population coincides with certain distribution (approximately log-normal). Also only a few researchers have investigated the filtration process of polydisperse particles by fibrous filter[12-15]. However, they only focused on the capture efficiency of clean fiber or the mass loading which depends on the filtration condition was artificially assumed. Kim[12, 14], Jung[13] and Song[15] studied the capture process of polydisperse particles (the particle size distribution is assumed as lognormal) through solving the k-order moment of particle size distribution function combining the existing formulas for calculating efficiency. Actually, the filtration process is an unsteady state, and the growth of dendritic structure, which depends on the initial particle distribution and fiber arrangement, influences the capture efficiency and pressure drop. Gaining insight into the dynamic evolution of the capture efficiency, pressure drop, deposition pattern, local porosity is very helpful to the design and optimization of filters.

In this work, the unsteady filtration process of polydisperse and fine particles was simulated using the Lattice Boltzmann-Cellular Automata (LB-CA) probabilistic model. The LB-CA model was successfully applied to simulate the steady and unsteady filtration processes of single fiber or a row of cylindrical fibers[16] and multi-fibers[17] in a laminar flow normal to their axes by our group. In addition, the dynamic process of dendritic structure growth and the dependence of the porosity of dendritic structure on the deposited particles were also investigated in this paper.

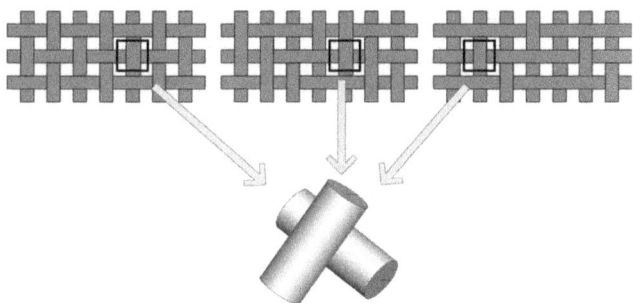

Figure 1 *Weaving patterns of filter cloth*

2. METHODOLOGIES

2.1 The Lattice-Boltzmann model for fluid flow

In the LB method, it is assumed that fluid consists of microscopic fictitious fluid-particles. The state of each grid is presented by distribution function $f_i(x,t)$ which indicates the probability density of the fictitious fluid-particles that locate on lattice x with the velocity c_i at time t. The fictitious fluid-particles on regular lattices have to experience two sequential sub-steps as follows:

$$\text{Collision: } f_i(x,t)' = f_i(x,t) + \Omega_i \tag{1}$$

$$\text{Streaming: } f_i(x+c_i \cdot \Delta t, t + \Delta t) = f_i(x,t)' \tag{2}$$

where Ω_i is collision operator. By combining two equations and introducing Bhatnagar-Gross-Krook (BGK single-relaxation) collision operator[18], the evolution equation of fluid system is obtained[19]:

$$f_i(x+c_i \cdot \Delta t, t + \Delta t) - f_i(x,t) = [f_i^{eq}(x,t) - f_i(x,t)]/\tau \tag{3}$$

where Δt is time step, τ is the dimensionless relaxation time, and f_i^{eq} is the equilibrium distribution function. For simplicity and without loss of generality, we choose the three-dimensional square lattice with fifteen velocities c_i ($D3Q15$ model, see Fig. 2, i runs from 0 to 14):

$$c_i = \begin{pmatrix} 0 & 1 & -1 & 0 & 0 & 0 & 0 & 1 & -1 & 1 & -1 & 1 & -1 & 1 & -1 \\ 0 & 0 & 0 & 1 & -1 & 0 & 0 & 1 & -1 & 1 & -1 & -1 & 1 & -1 & 1 \\ 0 & 0 & 0 & 0 & 0 & 1 & -1 & 1 & -1 & -1 & 1 & 1 & -1 & -1 & 1 \end{pmatrix} \tag{4}$$

Then the equilibrium distribution function can be calculated in the $D2Q9$ model as following:

$$f_i^{eq} = \rho \alpha_i [1 + \frac{\mathbf{c}_i \cdot \mathbf{u}}{c_s^2} + \frac{1}{2}(\frac{\mathbf{c}_i \cdot \mathbf{u}}{c_s^2})^2 - \frac{\mathbf{u}^2}{2c_s^2}] \tag{5}$$

where $\mathbf{u}$ is the macroscopic velocity of local fluid, ρ is the fluid density, α_i is the weight coefficient related to the model with $\alpha_0 = 2/9$, $\alpha_i = 1/9$ ($i=1\sim6$) and $\alpha_i = 1/72$ ($i=7\sim14$) in the D3Q15 model, c_s is the local sound speed, $c_s = \sqrt{3}c/3$ and $c = \Delta x / \Delta t$.

The macroscopic quantities of flow fields can be derived from statistics of distribution function, and the macroscopic density ρ and momentum $\rho\mathbf{u}$ are calculated as following:

$$\rho = \sum_{i=0}^{Q-1} f_i \ , \quad \rho\mathbf{u} = \sum_{i=0}^{Q-1} f_i \mathbf{c}_i \tag{6}$$

The equations for calculating the fluid viscosity and pressure are given by:

$$\begin{cases} v = \dfrac{c_s^2}{2}(2\tau - 1) \cdot \Delta t \\ P = \rho c_s^2 \end{cases} \tag{7}$$

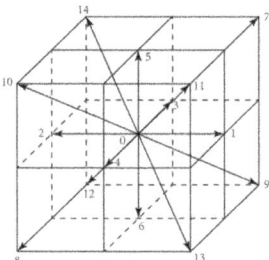

Figure 2 *D2Q15 model*

2.2 Cellular-automata (CA) probabilistic model for particle movement

In the existing CA model for particle motion[20, 21], the particle velocity is same as the local fluid velocity and the time step ratio of particle transport and fluid transport was considered as an empirical constant. It is difficult to quantitatively deal with the exact drag force and other external forces acting on the particles and describe the particle trajectory, the interphase interactions between fluid and particles using the existing model. Factually, the available LB–CA method in open literatures is only on the level of qualitative simulation. Wang et al. [16] established the quantitative model (see Figure 3), in which the transport probabilities of a particle to neighboring lattices depend on the combined effect of drag forces from fluid and Brownian diffusion[22]. The particle motion is governed by the following equation:

$$\frac{d\mathbf{u}_p}{dt} = \mathbf{F}_D + \mathbf{F}_B = \frac{\mathbf{u} - \mathbf{u}_p}{\tau_p} + \varsigma\sqrt{\frac{216\mu k_B T}{\pi \rho_p^2 d_p^5 \Delta t}} \tag{8}$$

$$\frac{d\mathbf{x}_p}{dt} = \mathbf{u}_p \tag{9}$$

where $\mathbf{u}_p$ is particle velocity, τ_p is relaxation time of particle, $\tau_p = \rho_p d_p^2/(18\mu)$, μ is dynamic viscosity of gas, $\mathbf{F}_D$ is drag force, $\mathbf{F}_B$ is Brownian force, ς is a zero-mean unit-variance independent Gaussian random number, d_p is particle diameter, k_B is Boltzmann constant, T is temperature of gas.

The particle velocity and displacement can be explicitly calculated through integration of the motion equations over time t successively:

$$\mathbf{u}_p^{n+1} = \mathbf{u}_p^n \cdot \exp(-\frac{\Delta t}{\tau_p}) + (\mathbf{u}_f + \mathbf{F}_B \cdot \tau_p) \cdot (1 - \exp(-\frac{\Delta t}{\tau_p})) \tag{10}$$

$$\mathbf{x}_p^{n+1} = \mathbf{x}_p^n + (\mathbf{u}_p^n - \mathbf{u}_f)(1 - \exp(-\frac{\Delta t}{\tau_p})) + \mathbf{u}_f \cdot \Delta t + (\Delta t + (1 - \exp(-\frac{\Delta t}{\tau_p}) \cdot \tau_p)) \cdot \mathbf{F}_B \cdot \tau_p \tag{11}$$

Superscript n and $n+1$ indicate present and next moment respectively. Then, the actual displacement of particle within Δt is obtained: $\Delta \mathbf{x}_p = \mathbf{x}_p^{n+1} - \mathbf{x}_p^n$.

In the CA probabilistic model, particles move along the regular lattices (same as the fictitious fluid-particles in the LB method) with certain probability p_i (Figure 3), which is proportional to the projection of its displacement on the lattice direction i:

$$p_i = \max(0, \frac{\Delta \mathbf{x}_p \cdot \mathbf{e}_i}{\Delta x}), (i = 1 - 6) \tag{12}$$

where $\mathbf{e}_i$ is the discrete velocity of fluid particles. Finally, the particle position after each time step Δt is determined as follows:

$$\mathbf{x}_p^{n+1} = \mathbf{x}_p^n + \mu_1 \mathbf{e}_1 + \mu_2 \mathbf{e}_2 + \mu_3 \mathbf{e}_3 + \mu_4 \mathbf{e}_4 + \mu_5 \mathbf{e}_5 + \mu_6 \mathbf{e}_6 \tag{13}$$

where μ_i is a Boolean variable and equal to 1 with probability p_i.

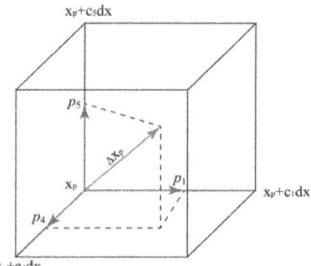

Figure 3 *Rule for particle transport in three-dimensional CA model*

3. RESULTS AND DISCUSSION

3.1 Deposition patterns of polydisperse particles

During the filtration process, the shape of collection medium (containing fibers and deposited particles) changes continuously (see Figure 4). The LB method, which is at advantage of easily dealing with complex boundary conditions, is used to simulate the gas flow field. This paper started from the clean fiber filtration process of particles having two different size distributions (condition 1 and condition 2, see Figure 5). For different filtration parameters (*Pe*, *R* and *St*), condition 1(left, average diameter is 1µm) is dominated by Brownian diffusion, interception and inertial impaction simultaneously, while condition 2(right, average diameter is 0.35µm) by diffusional mechanism only. Figure 5 also shows the particle size distribution at the outlet. It can be found that at the outlet the concentration of particles with diameter of about 1µm almost remains unchanged under condition 1. However, the larger or smaller particles are obviously captured. While under condition 2, the particle concentration at outlet is much smaller than the inlet. This is because the small particles with strong Brownian diffusivity are captured easily. Nevertheless, the intermediate particles (0.35µm, dominated by interception mechanism) are hard to be captured.

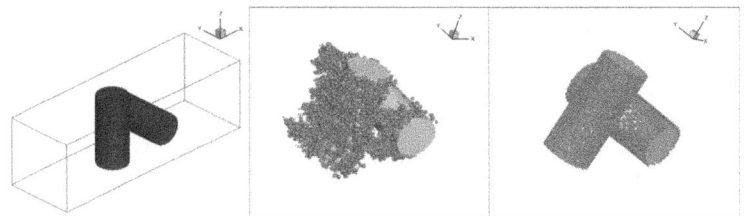

Figure 4 *Computational domain (left), deposition patterns due to interception (middle) and Brownian diffusion (right).*

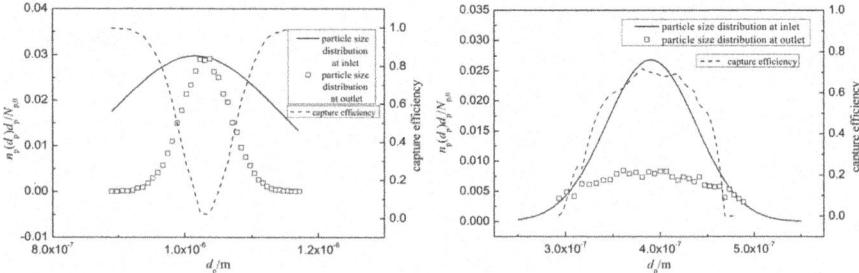

Figure 5 *Particle diameter distributions of inlet and outlet for clean fibers (left: condition 1; right: condition 2)*

However, during the dust loaded filtration process, particles often stick on the fiber surface which forms the dentritic structure (Figure 4, middle for deposition patterns in the interceptive regime and right in the diffusional regime). Most of micron-sized particles deposit in the windward area because interception mechanism is dominated in the process, while submicron particles deposit uniformly on the surface of fibers due to Brownian diffusion.

Figure 6 (top-left) presents the variation of porosity distribution and average porosity over mass loading (m, kg/m^2) under condition 1. The two-peak distribution of porosity (ε) is observed, where the first peak ($0.4<\varepsilon<0.5$) is formed by the deposition of larger particles and the second ($0.8<\varepsilon<0.9$) by smaller particles. With respect to polydispersed particle size distribution, larger particles could easily fill with a lattice while a lattice can accommodate many smaller particles. Therefore, at the initial stage, the porosity is large. With the detritic structure growing bigger, the value of first peak increases while the second decreases. Figure 6 (top-right) presents the average porosity as a function of the mass loading in condition 1, which approximately shows an exponential relationship through data fitting, i.e., $\varepsilon=0.197\exp(-m/3.43\text{e-}4)+0.394$.

Figure 6 (bottom) also shows the evolutions of porosity distribution and average porosity in condition 2. When comes to the tiny particles, a lattice can contain many particles. So the deposition of submicron particles results in a single-peak porosity distribution and the peak value is close to 1. In addition, the average porosity changes with mass loading as a linear function: $\varepsilon=0.991-294.42m$.

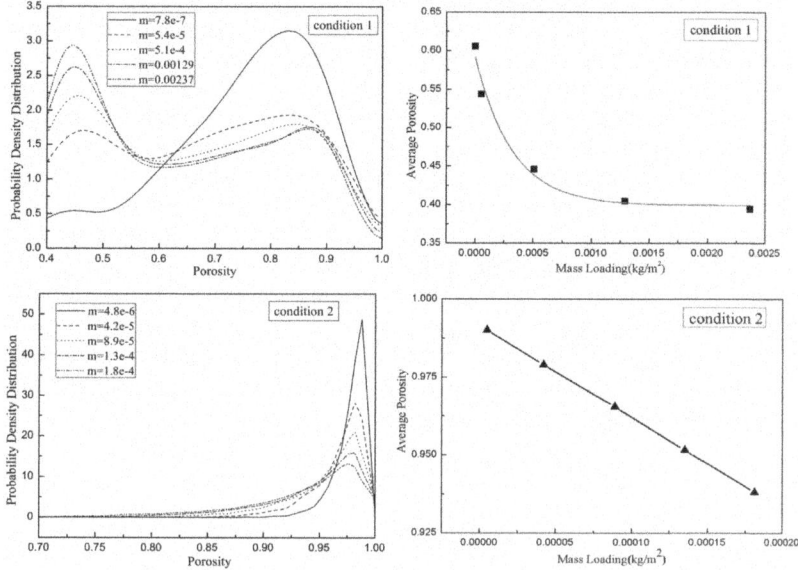

Figure 6 *PDF of porosity and average porosity evolution of dendritic structure under condition 1 and 2*

3.2 Filtration process of polydisperse particles in real filter

During the real filtration process, the particle size distribution is very complex and can not be expressed by a simple distribution function, moreover, different particles have different predominant mechanisms. Considering the flue gas from the coal-fired boiler, the particle size distribution is approximately characterized by a three-peak distribution[23], which can be viewed as a superposition of three lognormal distributions:

$$n_{\mathrm{p}}(d_{\mathrm{p}}) = \sum_{i=1}^{3} \frac{N_{\mathrm{p},i}}{\sqrt{2\pi} \ln \sigma_{\mathrm{pg},i}} \exp\left[-\frac{\ln^2(d_{\mathrm{p}}/d_{\mathrm{pg},i})}{2\ln^2 \sigma_{\mathrm{pg},i}}\right] \frac{1}{d_{\mathrm{p}}} \tag{14}$$

where $N_{\mathrm{p},i}$ is the number concentration of particles, $d_{\mathrm{pg},i}$ and $\sigma_{\mathrm{pg},i}$ are the particle's average diameter and standard deviation. Table 1 includes the values of all related parameters and figure 7 (left) demonstrates the initial particle size distribution at the inlet of computational domain.

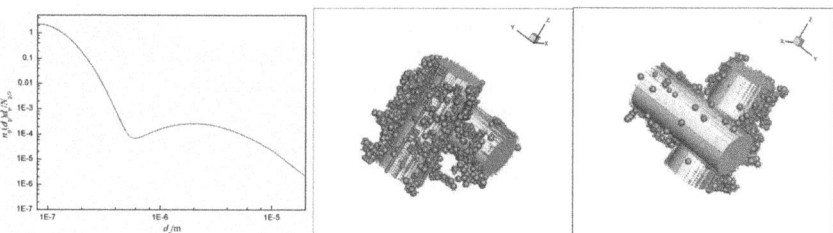

Figure 7 *Particle size distribution at inlet (left), deposition pattern on windward surface (middle) and leeward surface (right)*

Since the size distribution (Figure 7, left) spans over several orders of magnitude, three main mechanisms in real filtration process are all effective for different sized particles. Figure 7 (middle and right) shows that the deposition pattern of the three-peak distribution polydisperse particles. It is found that small particles can deposite everywhere on the fibers' surface (Figure 7, middle), while particles with larger diameter always deposite on the windward surface and there are a few large particles on the leeward surface (Figure 7, right). Obviously, the deposition pattern of large particles is similar to the previous condition 1, and that of small particles is similar to the condition 2. In addtion, the dentritic structure grows much faster for large particles than smaller particles.

Figure 8 demonstrates the probability density function (PDF) of porosity and average porosity over mass loading. In general, the PDF of porosity also shows a two-peak distribution during the whole filtration process, where the first peak is around 0.4 and second around 0.95. Moreover, with the mass loading increasing, the dentritic structure becomes more compact and the first peak increases while the second peak declines (although the two peak positions remain unchanged). Figure 8 (right) presents the average porosity as a function of the mass loading, which also shows an exponential relationship, i.e., $\varepsilon=0.394\exp(-m/0.00257)+0.527$.

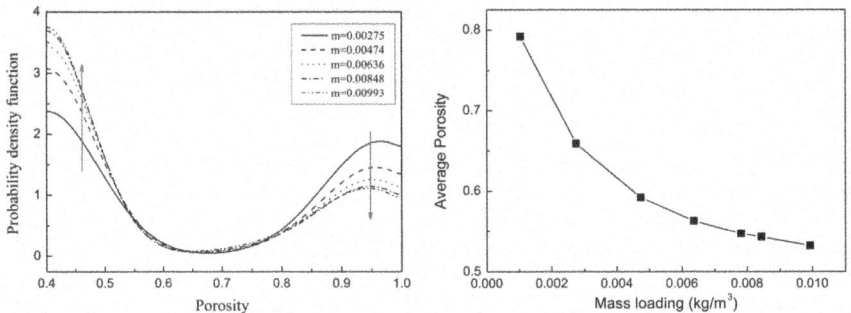

Figure 8 *PDF of porosity and average porosity evolution of dendritic structure in real filter*

4. CONCLUSIONS

Based on the real structure of fibrous filter and polydisperse particle distribution, the filtration process of 3D fibrous filter was investigated. Meanwhile, the growth process and structural characteristics of dendrite were explored in detail. Compared with existing model, the LB-CA model is good at considering the influence of fiber arrangement on filtration performance. It is found that, when diffusion mechanism is dominated, the porosity of dendritic structure obeys unimodal distribution and the average porosity is inversely proportional to mass loading. However, if there exists multi-mechanisms (such as small particles dominated by diffusion, medium particles dominated by interception), the porosity of dendritic structure follows a bimodal distribution and the average porosity is an exponential function of loaded mass. For particles deposition in real filter, two-peak distribution and exponential function are also observed.

ACKNOWLEDGEMENTS

Funding is provided by NSFC (51276077, 51021065), and MOST (2010CB227004).

References

1.   I. Stechkina and N. Fuchs, *Annals of Occupational Hygiene*, 1966, **9**, 59-64.
2.   A. Kirsch and N. Fuchs, *Annals of Occupational Hygiene*, 1967, **10**, 23-30.
3.   A. Kirsch and N. Fuchs, *Annals of Occupational Hygiene*, 1968, **11**, 299-304.
4.   I. Stechkina, A. Kirsch and N. Fuchs, *Annals of Occupational Hygiene*, 1969, **12**, 1-8.
5.   O. Filippova and D. Hänel, *Computers & Fluids*, 1997, **26**, 697-712.
6.   C. Kanaoka, H. Emi and T. Myojo, *Journal of Aerosol Science*, 1980, **11**, 377-389.
7.   Z. G. Liu and P. K. Wang, *Aerosol science and technology*, 1997, **26**, 313-325.
8.   G. Kasper, S. Schollmeier and J. Meyer, *Journal of Aerosol Science*, 2010, **41**, 1167-1182.
9.   T. Myojo, C. Kanaoka and H. Emi, *Journal of Aerosol Science*, 1984, **15**, 483-489.
10.   R. C. Brown, *Air filtration: an integrated approach to the theory and applications of fibrous filters*, Pergamon press New York, 1993.
11.   U. Lantermann and D. Hänel, *Computers & Fluids*, 2007, **36**, 407-422.
12.   H. Kim, S. Kwon, Y. Park and K. Lee, *Filtration+ Separation*, 2000, **37**, 37-42.
13.   C. H. Jung, H. S. Park and Y. P. Kim, *Environmental Progress & Sustainable Energy*, 2012, **31**, 397-406.
14.   S. Kwon, H. Kim and K. Lee, *Aerosol Science & Technology*, 2002, **36**, 742-747.
15.   C. Song and H. Park, *Powder technology*, 2006, **170**, 64-70.
16.   H. Wang, H. Zhao, Z. Guo and C. Zheng, *Powder technology*, 2012, **227**, 111-122.
17.   H. Wang, H. Zhao, K. Wang, Y. He and C. Zheng, *Journal of Aerosol Science*, 2013.
18.   P. L. Bhatnagar, E. P. Gross and M. Krook, *Physical review*, 1954, **94**, 511.
19.   Y. Qian, D. d'Humières and P. Lallemand, *EPL (Europhysics Letters)*, 1992, **17**, 479.
20.   A. Masselot and B. Chopard, *EPL (Europhysics Letters)*, 1998, **42**, 259.
21.   R. Przekop, A. Moskal and L. Gradoń, *Journal of Aerosol Science*, 2003, **34**, 133-147.
22.   B. Maze, H. Vahedi Tafreshi, Q. Wang and B. Pourdeyhimi, *Journal of Aerosol Science*, 2007, **38**, 550-571.
23.   W. S. Seames, *Fuel Processing Technology*, 2003, **81**, 109-125.

STUDY ON DEFOCUSED IMAGE PROCESSING METHOD FOR PARTICLE SIZE MEASUREMENT

J R Hu, W Zhou and X S Cai

Institute of Particle & Two-phase Flow Measurement, University of Shanghai for Science and Technology, Shanghai 200093, China

1 INTRODUCTION

The particle size and size distribution are vital components in the measurement of particle or particulate multiphase flow which existing widely in various industrial areas such as industrial fume monitoring, particle size measurement based on powder preparation and flow control in pipeline transportation. However, defocused particle images are apt to cover the true information of particles during online measurement based on image processing method. The particle sizes and size distributions are overwhelmingly influenced by the phenomena of defocus. Effective processing methods are needed to deal with these images.

The online measurement with image processing method requests the images in preferable quality that the image contrast of light and shade, clarity degree and whether there existing motion blur. The particles are not guaranteed to be totally in focus during the actual operation, blurred particles will inevitably make measurement results deviate from real value. The defocused image processing method proposed by this paper includes image pre-processing, boundary detection of blurred particles, and a criterion introduced to decide whether to count or not a particle given its degree of blur which aims at increasing recognition accuracy and recognition efficiency in size measurement of defocus blurred particles. Experiments are carried out to validate the proposed image processing method.

Consistently, in order to extend the measurement range, restoration in image processing for defocus blurred objectives is used. However, the particles with oversize defocusing amount are reluctant to be recovered on account of the perspective principle that the farther one objective is, the smaller the image will be, even with the same magnification. As a solution, we depict and analyze the properties of these defocus blurred particles on the boundaries and put forward the estimation method to distinguish and remove them, and in the meantime, particles close to the focal plane are restored.

2 IMAGE PRE-PROCESSING

The processing begins with a normalization of the image. The intention of this phase is to make the light source look homogeneous. We obtain a uniform background in the image based on the top-hat transformation which separating the particles from the background.

2.1 Bilateral Filter

Bilateral filter is a nonlinear filter method combined with the spatial proximity of the image and the pixel values of a compromise deal with the similarity by taking simultaneously the spatial information and the gray similarity into account. It achieves edge-preserving and noise-reducing purposes. Bilateral filter have the comparative advantage of simple, non-iterative and localized over the Veiner filter and Gaussian filter which smoother the boundary of the objective.

In the bilateral filter, the output pixel value depends on the weighed array of the value of neighborhood pixels is defined as

$$h(i,j) = \frac{\sum_{k,l} f(k,l)w(i,j,k,l)}{\sum_{k,l} w(i,j,k,l)} \tag{1}$$

the weighting coefficient $w(i,j,k,l)$ depends on the domain of nucleus

$$d(i,j,k,l) = \exp(-\frac{(i-k)^2 + (j-l)^2}{2\sigma_d^2}) \tag{2}$$

and the range of nucleus

$$r(i,j,k,l) = \exp\left(-\frac{\|f(i,j)-f(k,l)\|^2}{2\sigma_r^2}\right) \tag{3}$$

The weighting coefficient $w(i,j,k,l)$ can be expressed as the product of $d(i,j,k,l)$ and $r(i,j,k,l)$

$$w(i,j,k,l) = \exp\left(\frac{(i-k)^2+(j-l)^2}{2\sigma_d^2} - \frac{\|f(i,j)-f(k,l)\|^2}{2\sigma_r^2}\right) \tag{4}$$

The bilateral filter takes differences in the spatial domain and range into account simultaneously while the Gaussian filter and α average filter consider either spatial domain or range difference. The bilateral filter achieves great result in edge preservation and noise reduction as shown in Figure 1.

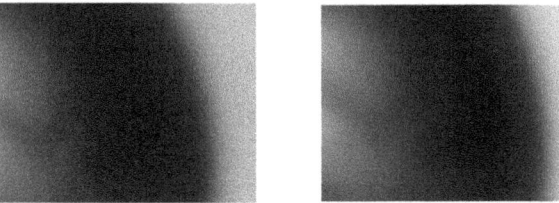

Figure 1 *Images before and after bilateral filtering*

3 DEFOCUSED IMAGE PROCESSING

3.1 Characteristics of Blurred Boundaries

The fact that the objects in the photo considered to be in focus or out of focus depends on whether the edges both inside an object and between objects are clear or not. When it comes to the particle size measurement, we are supposed to learn the information of the boundaries for the particles. The outlines of particles contain serviceable information to be available in the procedure of image processing that helps to reduce the volume of data calculated during image analysis. Unfortunately there has been no precise mathematical definition widely accepted on the edge thing, while the descriptive definition claims that it would be the border between two areas with different gray level, and inside each region, gray values are agreed. This special region between the object and the background is called the transition region: (1) it is a region with areas not equivalent to zero and distinguishing the objects from the background [1]; (2) the values of gray level in the transition region is also between those of objects and the background [2]; (3) the gray gradients in the transition region are higher than the gradients in other area, and the regional distribution of gray scale levels are richer [3]. It is generally considered, changes of gray levels in the peripheral direction of a circular pattern are relatively gentle, and on the contrary, it changes more intensely in the normal direction of the edge in gray scales. On the purpose of acquiring the gray scale characteristics in respect of the edge, Figure 1 compares the gray level transformation on the edge of a spherical particle in three different conditions from being focused, slightly unfocused and apparently out of focal field. The gray level curve of focused particle on its edge is approximately straight with large slope in Figure 1(a) and the gray gradients have two steeples. The difference value between two maximum of the gradient is equal to the diameter of the blurred particle [4]. In Figure 1(b), the curve corresponding to the particle edge becomes much more slant and (c) has got a smoother curve according to a blurred image boundary. These photos are taken of the transparent glass bead sample with light aperture around its centroid.

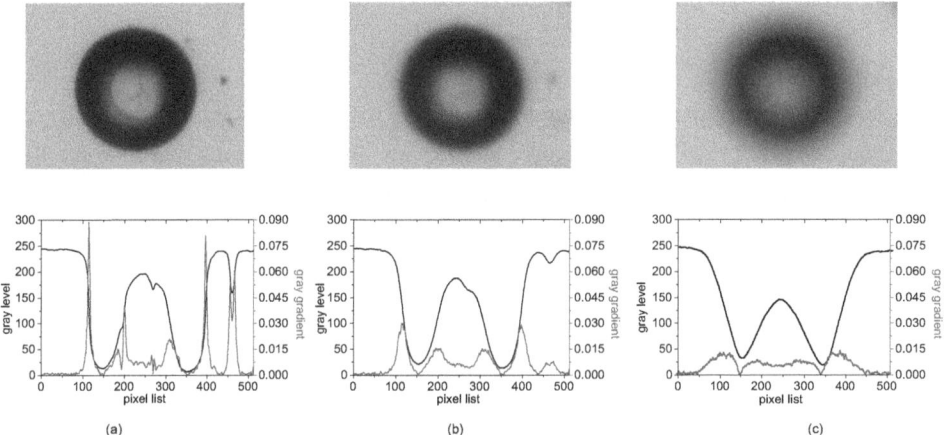

Figure 2 *Spherical particle images in different defocus degree and their corresponding gray scale and gray gradient curves*

In terms of the transformation curves above the edge of the particle in Figure 1(c), its small slope as well as the subtle variation of the gradient explains it is out of the focal plane. And the maximum values of gradient are confused. Obviously, the greater the defocus degree of one particle, the slower the edge of gray scale changes and the smoother the gradient goes, that is to say, the range of gray variation becomes wider which arouse the difficulties to estimate the actual position of the boundaries. As a pivotal variant, the gradients on the blurred boundaries are treated to be the foundation for separating defocused particles from others in the image.

3.2 Calculation of Gray Scale Gradient on the Boundaries

The image is digitized in computer for storage as the entire image can be replaced by a matrix. In the matrix every element contains the information of gray scale and location which represented by coordinate values of pixels in the image. Hence the image processing is to do numerical computation to each element of the matrix [5].

P is an m × n size matrix corresponding to an image containing 256 gray levels from 0 to 255, and $p(x_i, y_j)$ describes every element's gray scale value, so the matrix can be expressed as below.

$$P(x, y) = \begin{bmatrix} p(x_1, y_1) & p(x_1, y_2) & \cdots & p(x_1, y_n) \\ p(x_2, y_1) & p(x_2, y_2) & \cdots & p(x_2, y_n) \\ \vdots & \vdots & \vdots & \vdots \\ p(x_m, y_1) & p(x_m, y_2) & \cdots & p(x_m, y_n) \end{bmatrix} \quad (5)$$

If a pixel in the image falls on the boundary of the particle, then it will be a neighborhood with gray level changes. In order to obtain accurate description of the transformation characteristics, we introduce the gray gradient.

Supposing the gray scale field is $p(x, y)$, and the gradient column vector is

$$\text{grad } p(x, y) = \begin{bmatrix} \dfrac{\partial p}{\partial x} & \dfrac{\partial p}{\partial y} \end{bmatrix}^{\mathrm{T}} \quad (6)$$

The module for the gradient vector is

$$|\text{grad} p(x, y)| = \sqrt{(\partial p / \partial x)^2 + (\partial p / \partial y)^2} \quad (7)$$

Every pixel in the image is substituted for the module of the gray gradient, and then the image being processed is converted to a gradient matrix. Assume the processed image is $p(x, y)$, the gradient matrix is

$$g(x, y) = |\text{grad } p(x, y)| \tag{8}$$

Since $g(x_i, y_i)$ represents the gray gradient of any pixels on the boundary of the particle, calculate the statistical average of gradient magnitude and it is express as

$$bmgrad = \left[\sum_{i=1}^{u} g(x_i, y_i) \right] / u \tag{9}$$

In the equation, u is the quantity of pixels on the boundary of the particle.

3.4 Criterion

Define bmgrad as the value of particle's boundary gradient which presents the mean value of gray gradients on the particle boundary. Setting a criterion, the particles are in the state of defocus blur as if the boundary gradient is smaller than the criterion, and on the other side, a particle is to be preserved as its boundary gradient being higher than the threshold value. Yet the threshold does not absolutely distinguish the focus and defocus particles, in the reason that there is a deviation. In the particle size measurement applying to industrial, permissible error is acceptable. In conclusion, the particles are omitted when the gradient being under the criterion while the others are going to be restored and recorded to enhance the coefficient of utilization for the measurement results.

The determination of criterion is crucial in this method. Firstly, binarization is implemented on the particle image on the basis of Ostu segmentation (as shown in Figure 5(c)) and the area is calculated inward the boundary which is approximately within the peak region. And then regard the diameter of focused particle image as the real value expressed as d_0, and those in other occasions are called d whose deviation is calculated by the equation below

$$e = (d - d_0) / d_0 \times 100\% \tag{10}$$

4 EXPERIMENTS

As a result of perspective, the objective in front of the lens has different sizes when it moves back and forward along the optical axis of the camera. The objects within the field depth of telecentric lens sustain the same sizes and appear to be distinct with human eyes when displayed in the computer screen. Taking the advantage of telecentric lens that the particle's size remains the same when it is within the depth of field (DOF), the absolutely clear condition of a particle can be easily found.

4.1 Experimental Setup

The experimental system is shown in Figure 3. The telecentric continuous zoom lens (VSZ-0745CO) is connected with the CCD camera (UI-5240CP). To capture images in different defocus degrees, the position of sample of glass beads is adjusted by the micrometer spiral in the axis direction of the lens. The LED light source is adopted for backlighting. And the data acquisition of CCD is transmitted through the cable to the computer.

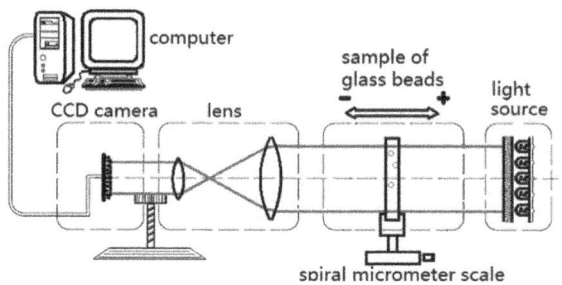

Figure 3 *The experimental setup*

4.2 Calibration

Calibrate the system under 1.5-times magnification by shooting a picture of the particles in the DOF. The length of cross-line graticule denoted as δd covers pixels in the number of δm, so that every pixel in the image is equivalent to the length of $\delta l = \delta d / \delta m = 3.52 \, \mu m/ \text{pixel}$. The binary processing procedure is going to be carried on to the original image after filtering method. We apply the function of 'regionprops' in MATLAB to recall the value of the area s_p embraced in the particle boundary. Assuming that the particle is spherical, therefore, the diameter could be expressed as $d_p = \sqrt{4s_p / \pi} = 219.0419$. We convert pixel diameter into the actual diameter by the equation $d = d_p \times \delta l = 771.0276 \mu m$.

4.3 Determination of the Criterion

Every movement of the spiral micrometer scale is taken at 0.5mm. The DOF of the telecentric lens is 1mm under 1.5-times enlargement ratio, there is at least one set of data enable to be gained under focusing condition.

As the glass beads sample moves from the near to the distant from the lens, their images are shot and stored in computer processor. The data of particles pixels are subsequently converted to diameters and their deviation as well as corresponding gray gradients changing with the distance from the focal plane are reported in Figure 4 which have been operated by the programs written in MATLAB. Supposing that the deviation for particle measurement is ±7.5%, the particles whose diameter deviation less than the given value are selected and the smallest gray gradients among them is picked as the criterion we need. Finally, restore the particles whose boundary gradients are above the threshold and omit the other particles not meeting the requirements of the criterion. The data in Figure 4 indicate that within the acceptable deviation, that the criterion is selected as 0.02295.

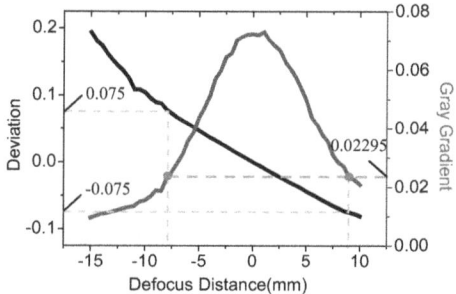

Figure 4 *Spherical particle diameter and deviation calculated from boundary gradient in different defocus distances*

5 RESULTS AND CONCLUSION

We record particles in the focus and restore the slight defocusing ones, however delete the apparently fuzzy particles when processing an image including all kind of particles whether in or out of the focal plane. And Figure 5 reveals the course of the image processing method effectuated on MATLAB software involving normalization, filtering, binarization and blur particles omitting.

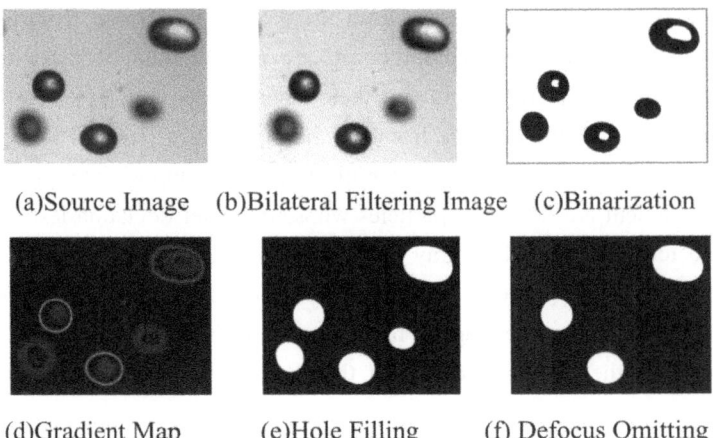

(a)Source Image   (b)Bilateral Filtering Image   (c)Binarization

(d)Gradient Map   (e)Hole Filling   (f) Defocus Omitting

Figure 5 *Defocused Image Processing*

This paper studies on the image processing method in particle size measurement by analyzing the defocus blurred phenomena in on-line measurement and presenting solutions. An estimation method for the gradient threshold is proposed to distinguish the particles of different blur occasion by given a criteria depending on the diameter and gray gradient at varied distance from the focal plane. Satisfying results can be achieved for particles of 600 µm to 800 µm diameters based on the proposed method.

On-line measurement of particle size characteristics and other information on the basis of image processing orders certain requirements on the acquires images, such as there having obvious contrasts in the image to distinguish the objects and background, and minimizing image noise caused by other environmental factors, also, the imaging pixels of per project are suggested not be less than 20 pixels. Therefore, when applied to online measurement the method demands severe control of acquisition environment. When selecting the gradient threshold, considerable specimens similar to the measured particles had better to be taken and the image acquisition environment is proximate to certain occasion of the measurement spot.

ACKNOWLEDGEMENT

The authors acknowledge the support from Shanghai Universities Young Teachers Training and Funding Scheme (Grant No. SLG12002); the National Natural Science Foundation of China (Grant No. 51206112); Shanghai Natural Science Foundation (Grant No. 12ZR1446900)

References

1. Gerberands J J. Segmentation of noise images [D]. The Netherlands: Delft University, 1988
2. Zhang Y J, Gerberads J J. Transition region determination based thresholding [J]. Patter Recognition Letters, 1991, 12(1): 13-23
3. Yan Chengxin, Sang Nong, Zhang Tianxu. Degree-based image transition region extraction and its segmentation [J]. Journal of Huazhong University of Science and Technology; Natural Science, 2004, 32(10): 1-3
4. CHENG Linhu,CAI Xiaoshu,ZHOU Wu. Particle sizing from defocus image of spherical particles by image transition region gradient method [J]. CIESC Journal, 2012, 63(12): 3832-3838.
5. BAI Jingfeng, ZHAO Xuezeng, QIANG Xifu, YANG Yanzhu. Edge Detection Based on Fuzzy Gradient Method [J]. Control and Decision, 2001, 16(3): 351-354

PREPARATION AND PERFORMANCE OF PANI/MWNT COMPOSITE FILMS ANODE FOR MICROBIAL FUEL CELL

X. Y. Wang, H. B. He, C. C. Zheng and Q. J Guo

College of Chemical Engineering, Qingdao University of Science and Technology, Key Laboratory of Clean Chemical Processing Engineering of Shandong Province, Qingdao 266042, Shandong, China

1   INTRODUCTION

Biologists estimated that Earth's land could produce 100~125 billion tons of biomass, while the ocean produces 50 billion tons of biomass each year. Biomass energy occupies an important position in the whole energy system. Sewage and industrial organic wastewater are one form of biomass, which were sent to a sewage treatment plant that consumed large amounts of energy due to the enormous equipments and technical requirements etc. Anaerobic treatment technologies provide a potential for cut down treatment and operating costs.[1] Microbial fuel cells, which are new technologies to process wastewater while producing electricity simultaneously, [2-4] can solve these problems. Moreover, MFCs are suitable for using various forms of biodegradable organic wastewater, even sludge, etc. [5-6]

Up to dates, the main challenge is to obtain sufficient power density and the optimal voltage as large as possible. Many investigators have attempted to increase the MFC voltage and power output by the selecting and cultivation of high quality microbial organisms, [7-9] by optimization of the electrode material and its surface structure, [10-14] by seeking inexpensive and efficient catalyst of cathode, by improving the configuration of MFC. However, the maximum voltage will never exceed the theoretical open circuit voltage (OCV), for instance, the open circuit voltage of the air cathodes MFC using the acetate as the substrate was 1.105 V even neglecting the internal resistance losses. [15]

Anode materials and their structure are key factors for successful application performance of MFCs. The specific performances consist of its effects on bacteria adsorption, electron transfer, matrix oxidation, etc. A large number of materials have been applied to the microbial fuel cells. Therefore, the use of carbon-based electrodes in paper, foam forms, and reticulated vitrified carbon (RVC) for the MFC anode is very common. A large variety of graphite-based electrodes such as graphite rods (felt, particles, sheet, etc) are also widely used in MFC. Furthermore, much attention has paid on conductive polymers, metals and metal coatings, non-metallic treatments and modifications of anode materials, have a certain research up to now. Particularly the use of carbon brush and graphite brush, which have observable characters of the large specific surface area and porosity, greatly enhances the electrogenesis capacity of a microbial fuel cell. Logan et al. [13] compared large graphite brush anodes and random fibers in single-chamber bottle MFCs and obtained that large graphite brush anodes MFC has a higher power density than carbon paper anode MFC. They also found that fiber clumping was a factor in the performance of this system. In addition, ammonia treatment of carbon-cloth anodes is a most successful method to enhance the power output from MFC. [12] Although the surface charge increased by 10 times, ammonia treatment process significantly improves the cost. Tsai et al. [16] used MWNT-modified carbon cloth as the anode, and investigated without the Pt catalyst and PEM membrane. The maximum power density is 65 mW·m^{-2}, and the voltage is just about 200 mV, which is far away to sufficiently maintain its running cost.

It is difficult to determine if a material performed better than another because of that different devices being employed. In addition, another important reason is the high internal resistance limits power generation, as well as increasing anode surface area may not appreciably affect power output .[15] Huang et al. [17] utilized different metals as the anode of a two-chamber MFC, got the conclusion that metals have a better anode performance, which could increase the transmission efficiency of electrons in the anode chamber, although metals have a bad stability, especially corrosive organic sewage is used as the matrix. Current densities using a graphite rod, graphite felt or graphite foam as anode were compared by Chaudhuri et al. ,[18] who found that the total accessible geometrical (projected) surface area caused the main difference of current densities, their further research was not reported. Huang et al. [19] compared carbon paper, graphite and carbon felt materials, and investigated the influence of five anodic characters such as the pore volume, surface area, aperture distribution, surface roughness and surface electric potential on the power generation performances of MFC. Li et al. [20] compared the performance of conventional carbon cloth anode and granular activated carbon anode in the dual-chamber MFC, obtained that large specific surface area is responsible for granular activated carbon anode MFC's better performance, but the particulate carbon is easy filled with the anode chamber, leading to plug and be difficult to clean up.

In this paper, liquid-solid fluidized bed is combined with single-chamber air cathode MFC reactor, constructing anaerobic fluidized bed microbial fuel cell (AFBMFC). Polyacrylonitrile / Multi-walled carbon nanotubes (PANI/MWNT) composite films anode was prepared in cyclic voltammetry (CV) method and applied to AFB-MFC. The surface structure, the electrochemical properties, and the prepared electrodes were investigated. The output voltage, the power density and the efficiency of sewage treatment for MFC were evaluated.

2 MATERIALS AND METHODS

2.1. Experimental Apparatus

As illustrated in Figure 1, AFBMFC was constructed by a plexiglas vessel, and the internal diameter and height of the fluidized bed anode chambers are 4 cm and 60 cm, respectively. It consists of a porous anode and a cathode, which is made of a carbon cloth (3.14 cm^2, 0.35 mg·cm^{-2} Pt) and coated with four diffusion layers, fixed onto one side of the chamber wall. 60 g fresh active carbon particles with an average diameter of 0.45~0.9 mm, bulk density 566 kg·m^{-3}, true density of 1150 kg·m^{-3}, porosity of 0.45 were fed in fluidized bed and were used as the carrier media for biofilm. The distributor was a porous glass plate with a thickness of 2 mm, a pore size of 2 mm, which has a fractional perforated area of 20%.

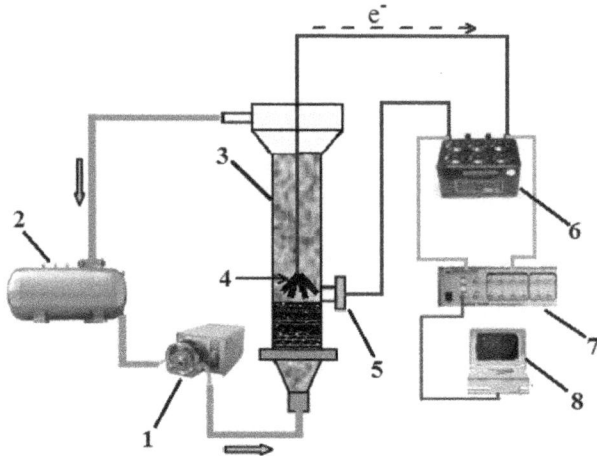

Figure 1 *The schematic diagram of anaerobic fluidized bed MFC*
1-peristaltic pump; 2-water storage tank; 3-fluidized anode chamber; 4-anode; 5-air cathode; 6-external resistance; 7-data acquisition system; 8-computer

2.2. MFC Inoculation and Operation

The AFBMFC was inoculated with 150 ml of anaerobic sludge collected from the Wastewater Treatment Plant, Qingdao, China. Mixed cultures are more suitable for the use of complex fuels such as wastewater, as single organisms generally metabolize quite a limited range of organic compounds. Wastewater, collected from restaurant wastewater of a university, was used as the fluidization liquid supplied with a peristaltic pump. The pH of the solution in the anode chamber was initially adjusted to around 7. The feed solution was replaced when the voltage dropped below 50 mV, representing one complete cycle of operation. A water bath was used for maintaining at the temperature of 30 °C. AFBMFC experiments were generally performed at 30 °C unless otherwise specified.[21]

2.3 Preparation of Anodes

2.3.1 Pretreatment of MWNT
MWNT has an excellent performance, but pretreatment is necessary to remove amorphous carbon, chopped, impurities, connect the corresponding functional groups, and to increase dispersed in acid solution. Performances of commercial MWNT used in this experiment are showed in Table 1.

Table 1 *Properties of multi-wall carbon nanotubes*

OD / (nm)	Length / (μm)	Purity / (wt%)	Ash / (wt%)	SSA / ($m^2 \cdot g^{-1}$)	EC / ($S \cdot cm^{-1}$)
8~15	~50	>95	<1.5	>233	>10^2

MWNT were ultrasonically dispersed in a 3:1 [25] concentrated sulfuric and nitric acid mixture for 6 hours (50 Hz), which is aimed at producing carboxylic acid groups at the defect sites and thereby improve solubility in acid solution. The mixture was set still for 1 h before diluting with a mass of de-ionized water. After that, the particles and agglomerates in large sizes were removed from the solution by centrifuging. The centrifugal liquid was carefully decanted in a glass beaker, filtered by vacuum pump, and washed with de-ionized water until the slurry with pH 6~7 was obtained. This slurry was washed several times with a small amount of ethanol and acetone, respectively, until the filtrate indicates colorless. Finally, the acid-treated MWNT was washed with de-ionized water again and dried in vacuum drying oven. The product was clean and stable dispersed in acid solution.

2.3.2 Preparation anodes by electrochemical method

PANI/MWNT film anodes: To take advantage of three electrodes system, CV method was utilized. The condition is hydrochloric acid as the proton acid, while its concentration is 1 $mol \cdot L^{-1}$ in the electrolyte. Meanwhile, the aniline concentration is 0.1 $mol \cdot L^{-1}$. Different scanning laps were used to prepare PANI/MWNT film anodes. After the preparation, the electrode surface acid was washed neatly with distilled water, then, dried more than 24 h, standby application. Prior to each electrochemical synthesis, the working electrode was not only carefully polished with abrasive paper and then washed with distilled water but also scanned and activated in 1 M sulfuric acid solution, by voltammetric sweep between -0.5 V and 1.5 V/SCE, at 50 $mV \cdot s^{-1}$. Scanning was finished until forming a stable CV curve.

2.4 Material Analysis and Calculations and Analyses

2.4.1 Material Analysis

In this paper, electrochemical methods such as CV and electrochemical impedance spectroscopy (EIS) were used to analyze electrochemical performances of anode materials. And SEM and TEM were used to characterize the electrode surface structure and morphology.

2.4.2 Calculations and Analyses

The electronical potential across the resistor was recorded every two minutes using a multimeter with a data acquisition system (USB1608FS, measurement computing Co., American). Polarization data were collected by changing the external resistance (varied from 30 Ω to 90 kΩ) by a variable resistor box during the stable power production stage of each batch experiment. The current, power density, resistant, COD removal rate (national standard: GB11914-89), coulombic efficiency were calculated as follows equations:

$$I = \frac{U}{R} \tag{1}$$

$$P = \frac{UI}{RA} = \frac{U^2}{RA} \tag{2}$$

$$U = -rI + U_{OCI} \tag{3}$$

$$B = \frac{C_0 - C_i}{C_0} \times 100\% \tag{4}$$

$$C_E = \frac{q}{q_{th}} \times 100\% = \frac{\int Idt}{(F \times b \times V \times \Delta COD)/M_{O_2}} \times 100\% \tag{5}$$

Where U (mV) is the voltage, P (mW·m^{-2}) is power density, R (Ω) is the external resistance, U$_{OCV}$ (mV) is open-circuit voltage, r (Ω) is internal resistance, and A (cm^2) is the geometric surface area of the anode electrode. T is reaction time (s), V is the liquid volume in the reactor (L), ΔCOD is the change in COD (kg·L^{-1}), M_{O_2} is the molecular weight of oxygen (32, g·mol^{-1}), I is current (A), F is Faraday's constant (96485, C·mol^{-1}), b is the mol number of electrons exchanged per mole of oxygen (4, mol e$^-$·mol^{-1})[22].

3   RESULTS AND ANALYSIS OF FILM ANODES AFBMFC

3.1 Influence of Polymerization Laps

Figure 2 is the CV curve of PANI/MWNT films prepared by electroplating 32 circles. At the first lap, the current promptly increased after the voltage is greater than 0.8 V, which is due to chain initiation of PANI. Then three redox peaks were emerged, and peak current increased with the increasing of polymerization laps. The redox peaks are diminished when the polymerization laps are more than a certain value, which is about 30. That the redox reaction is almost stopped can be responsible for the above phenomenon. In this experiment, different laps of 8, 16, 24 and 32 were chosen to assess the performance of AFBMFC.

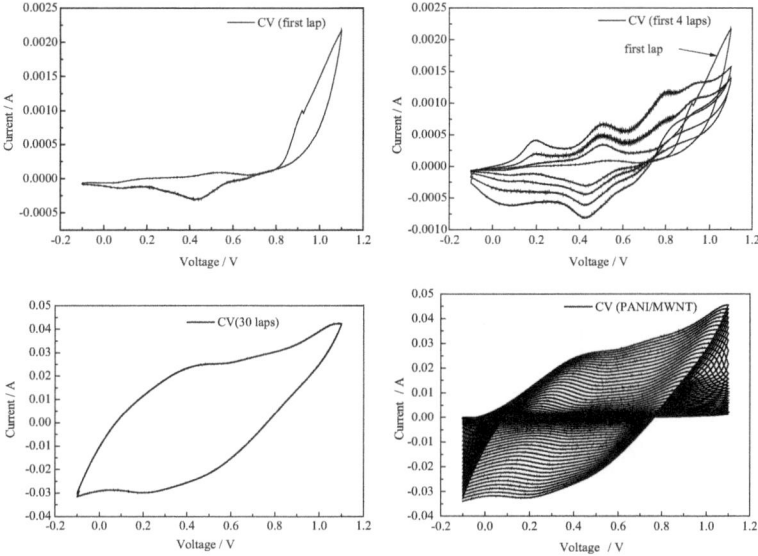

Figure 2 *Cyclic voltammgrams of PANI/MWNT films prepared by electroplating 32 circles*

3.2 Electrochemical Impedance Spectroscopy (EIS) Studies

Electrochemical impedance spectroscopy (EIS) measurements were carried out to compare the characteristics of charge transfer and ion transport in PANI/MWNT composite made from different laps. The test parameters are set as follows: the initiate frequency is 100000 Hz and terminate amplitude is 0.01 Hz and polarization amplitude is 100 mV vs. SCE. The results in Figure 3 show that single semicircles over the high frequency range, followed by short straight lines in the low-frequency region for all anodes. The equivalent circuit diagram of Figure 3 was applied to fit the measured EIS results, and the fitting results are showed in Figure 4.

As showed in Figure 4, the total resistance of all anodes gradually decreased with the polymerization laps, in spite of each part of the resistance also changes along with the polymerization laps. We can obtain that the electrochemical reactions resistance (Rct) as well as diffusion resistance (Ws) is decrescent while solution resistance (Rs) is nearly invariable. The Rct has primary responsibility to internal resistance. Interaction between PANI and MWNT promotes charge transfer in PAMI/MWNT composite. The MWNT have an obvious improvement effect for a faster charge transfer rate due to its favourable electrical conductivity. The result reveals that the film anodes not only improves the electrode conductivity and but also provides increases specific surface-area, possibly due to its specific nanostructure, which is benefit to host the bacteria and gain efficient degradation ability of wastewater.

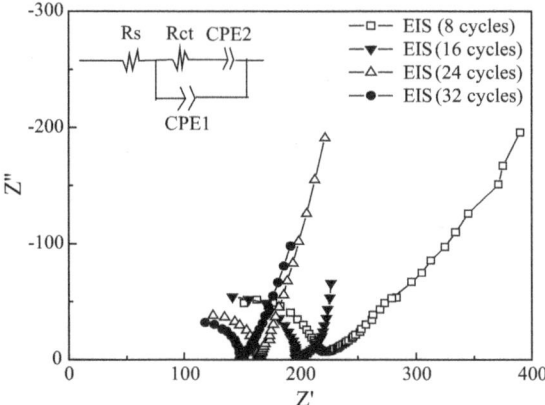

Figure 3 *EIS of anodes made in different cycles by CV method (Nyquist plot)*

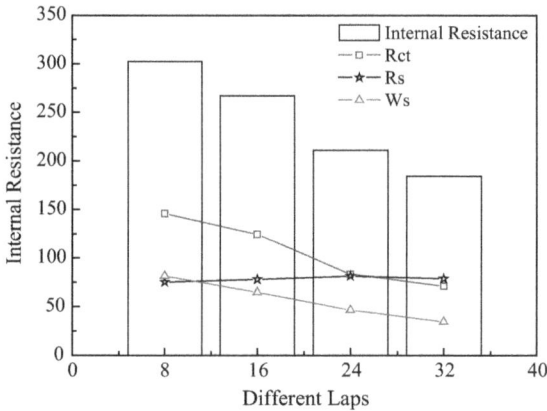

Figure 4 *The every part scattergram of resistance*

Considering the thick film could fall off from the electrode surface while it is dry, although preparation of anodes for cycling 32 laps could decrease the resistance. As a result, anode preparation should choose an appropriate lap, and the proper laps of 20~30 was chosen.

3.3 Performances of AFBMFC with PANI/MWNT Anodes

On the basis of EIS analysis of PANI/MWNT film anodes, the performances of anodes made from 8 and 24 laps were assessed in AFBMFC.

3.3.1 Maximum Voltage of AFBMFC in One Cycle

Figure 5 spreads the maximum voltage of AFBMFC in one cycle with PANI/MWNT film anodes made from 8 and 24 laps using the CV method. The PANI/MWNT anode (8 laps) AFBMFC has an initial stable voltage of 875mV approximately, which improves about 500 mV than the graphite rod as the anode. And the voltage is stable between 800 and 900 mV. Maximum voltage is 893.8 mV in the third day. The PANI/MWNT film anode (24 laps) AFBMFC has an initial stable voltage of about 850 mV, and has a stable voltage of 950 mV for four days after working 30 hours. Maximum output voltage is 967.7 mV, which increases 50 mV or so than PANI/MWNT anode made from potentiostatic method in previous researched [23]. In addition, after 150 hours, the output voltage decreased gradually. It is possibly ascribed to that the decreased of the organic matter.

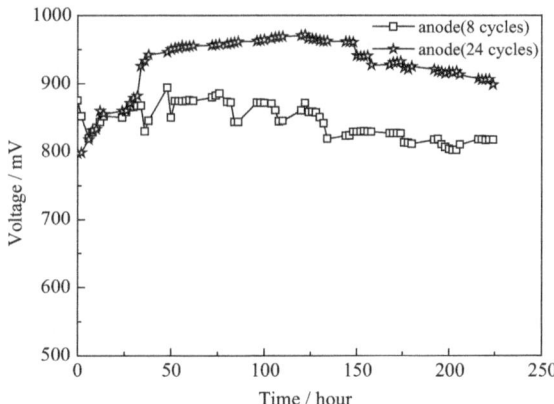

Figure 5 *Maximum voltage of AFBMFC in one cycle*

3.3.2 Polarization Curve and Power Density Curve

5 wt% MWNT in the PANI/MWNT film anodes prepared by CV method polymerizing 24 cycles were optimal, whose maximum output voltage was 967.7 mV as its power density was 286.63 mW·m^{-2}, as showed in Figure 6. Fitting the polarization curve according to equation (3), apparent resistance of 973.6 Ω was obtained, and which is responsible for the low voltage and power output.

Usually， the apparent resistance of MFC mainly includes three parts: ohm resistance, activation resistance and concentration differential resistance. The ohm resistance is mainly because that the transmission of electronic or ions is hampered by the electrolyte. And the activation resistance is caused by the activation reaction on the electrode surface. And yet the concentration difference resistance is due to the low diffusion speed of reactant to the electrode surface or the low spread speed of reaction products in the solution.

The AFBMFC, fluidized bed combined with MFC, could mix matrix fully and have a high-efficiency mass transfer. That is better to reduce the concentration difference resistance. Whereas, the fluidization interbody, coconut shell activated carbon, not closing contact to anodes could reduce the electronic recovery, and then the apparent resistance is large. Moreover, AFBMFC is a direct-air cathode single-chamber microbial fuel cell, and Liang et al.[24] found that the air cathode MFCs have bigger ohm resistance. Therefore, to reduce the ohm resistance and the activation resistance become the chief aspects to reduce the internal resistance of AFBMFC. The method of increasing the electrode area, improving the anode shape, enhancing ionic strength, optimizing the operation conditions, domestication highly active microorganisms etc can be used to reduce the internal resistance of AFBMFC.

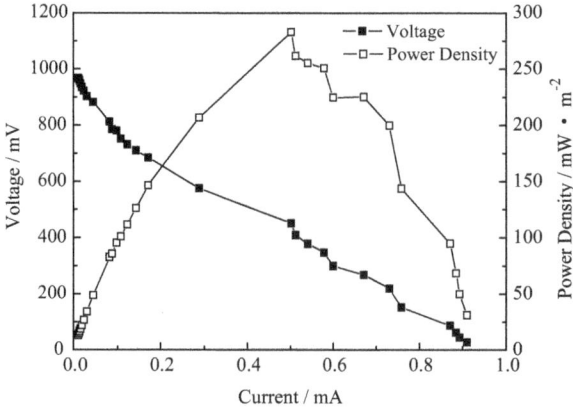

Figure 6 *Polarization curve and power density curve of CV method for PANI/MWNT film anode AFBMFC*

3.3.3 SEM and TEM

Scanning electron micrographs of PANI /MWNT film anodes made from CV method (5 wt% MWNT containing, 24 laps) are showed in Figure 7. The PANI/MWNT composite films have a networked-rod nanostructure, and its surface has many raised fold shapes, which increased the surface area and specific surface area, providing many growing points for microorganisms. The surface of used PANI/MWNT film anode formed membrane-shape material, which consists of microorganisms and their metabolin. Therefore, it is easy to transmit electrons and protons and so on, as well as to improve the degradation speed of organic matter in sewage.

Transmission electron microscopys of MWNT and PANI/MWNT composite film are showed in Figure 8. Results show that: the diameter of MWNT is about 8~15 nm and has a hollow tubular structure. While the PANI/MWNT presents that the outer layer is PANI and the inner layer is. The amorphous and lamellate outer PANI layer has a thickness of about 80~100 nm. There is Π combined key between PANI and MWNT, which increases the stability of PANI/MWNT composite films as well as averted the poisonousness to microbes of MWNT.

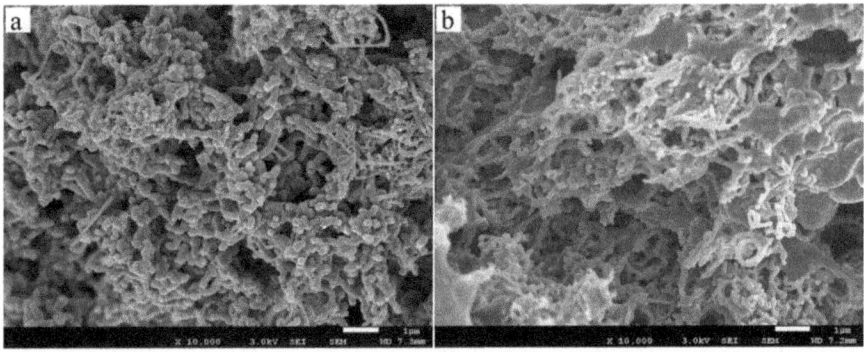

Figure 7 *SEM images of 5 wt% PANI/MWNT film anode made in cyclic voltammetry method (a: unused; b: used)*

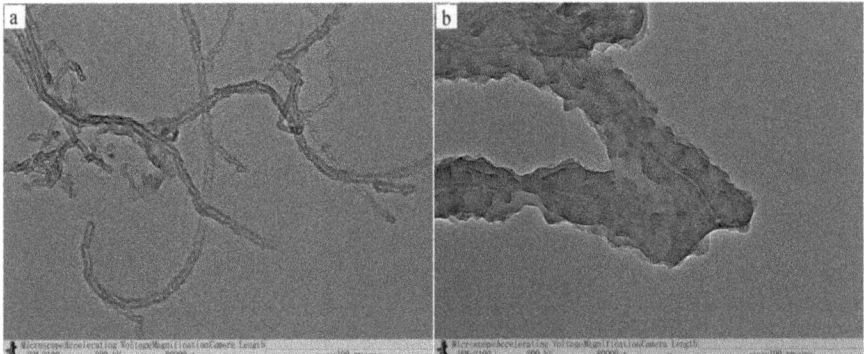

Figure 8 *TEM images of acidulated MWNTs and 5 wt% PANI/MWNT film anode made in cyclic voltammetry method (a: acidulated MWNT; b: PANI/MWNT film anode)*

3.3.4 COD and Electronics Recycling

The chemical oxygen demand (COD) was measured in one cycle of AFBMFC with PANI/MWNT film anodes (8 laps and 24 laps), and the results were shown in Figure 9. The COD removal rate is 92.1% and 95.3%, respectively. The results revealed that PANI/MWNT film anode made from 24 laps with CV method has better performance and increased the consuming of organic matter.

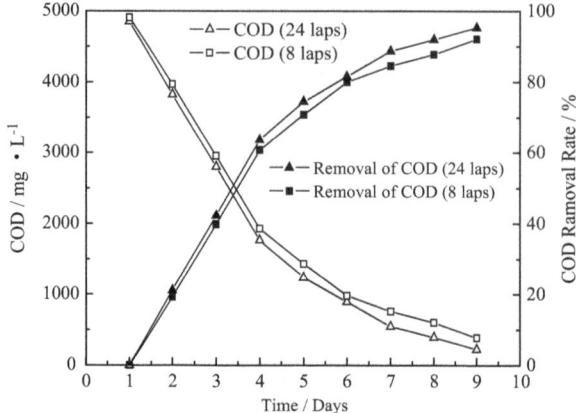

Figure 9 *Curves of COD when the film anode at scanning laps of eight and twenty-four*

Now, electron transport of MFC basically via three mechanisms: electron shuttling via self-produced mediators or chemical mediators, nanowires and cytochrome. The MWNT doping in film anodes facilitated the electronic recovery. This is mainly due to the increasing of the electrode surface area. However, the electronic recovery is very low with different anodes in AFBMFC. On the one hand, mixed bacterium were used in this experiment, that the ability of produce electrons was reduced because of the oxidation organic substrate by methanogen and thiogenic bacterium etc. On the other hand, the internal resistance is higher, further study was needed to reduce that.

4   CONCLUSIONS

(1) The appropriate number of polymerization cycles is 20~30 when graphite rod as the working electrode, CV method is used for the preparation of PANI/MWNT film.
(2) 5 wt% MWNT in the PANI/MWNT film anodes prepared by CV method polymerizing 24 cycles were optimal, whose maximum output voltage was 967.7 mV as its power density was 286.63 mW·m^{-2}. And the maximum voltage is approximately 950 mV for four days. The COD removal rate of the wastewater after a cycle is reached 95.3%.
(3) PANI/MWNT composite films anode made in cyclic voltammetry (CV) method is a high-efficiency anode to use for AFB-MFC.

ACKNOWLEDGMENTS

The research was supported by the Doctoral Program of Higher Education of China (20103719120004), the Natural Science Foundations of Shandong Province, China (ZR2013BM012), and the College science and technology plan of Shandong Province, China (J11LB62).

Notation

A	=	anode area (m^2)
B	=	removal of COD
b	=	mol number of electrons exchanged per mole of oxygen (4, mol $e^-\cdot mol^{-1}$)
C	=	1 C= electric quantity consist by $1.6*10^{19}$ electrons
C_E	=	coulomb's efficiency
C_0, C_i	=	chemical oxygen demand of waste water before and after the operation ($g\cdot L^{-1}$)
D	=	diameter of MWNT (nm)
F	=	Faraday's constant (96485 $C\cdot mol^{-1}$)
I	=	current (mA)
L	=	length of MWNT (μm)
M	=	amount of substance ($mol\cdot L^{-1}$)
M_{O_2}	=	molecular weight of oxygen (32 $g\cdot mol^{-1}$)
N	=	polymerization laps
P	=	output power density ($mW\cdot m^{-2}$)
Q	=	cathode sedimentary power (C)
q, q_{th}	=	harvestable electric quantity and theoretical maximum electric quantity (C)
R	=	load resistance (Ω)
r	=	internal resistance of MFC (Ω)
t	=	reaction time (s)
U	=	output voltage (mV)
U_{OCV}	=	open-circuit voltage (mV)
V	=	matrix volume (L)

References

1. H. Liu, B. E. Logan, *Environ. Sci. Technol.*, 2004, **38**, 4040.
2. M. C. Potter, *Proc. R. Soc. London, Ser. B.*, 1911, **84**, 260.
3. J. R. Rao, G. J. Richter, F. Von Sturm and E. Weidlich, *Bioelectrochem. Bioenerg.*,1976, **3**, 139.
4. B. E. Logan, J. M. Regan, *Environ. Sci. Technol.*, 2006, **40**, 5172.

5. K. Scott, I. Cotlarciuc, D. Hall, J. B. Lakeman and D. Browning, *J. Appl. Electrochem.*, 2008, **38**, 1313.
6. K. Rabaey, W. Verstraete, *Trends Biotechnol.*, 2005, **23**, 291.
7. S. K. Lower, M. F. Hochella and T. J. Beveridge, *Science.*, 2001, **292**,1360.
8. K. Rabaey, N. Boon, M. Hofte and W. Verstraete, *Environ. Sci. Technol.*, 2005, **39** , 3401.
9. M. Liu, J. Shao, B. Zhou, S. G. Zhou and N. I. Jinren, *Chin. J. Appl. Environ. Biol.*, 2010, **16**, 445.
10. Y. Zhao, S. Nakanishi, K. Watanabe and K. Hashimoto, *J. Biosci. Bioeng.*, 2011, **112**, 63.
11. C. Li, L. B. Zhang, L. L. Ding, H.Q. R and C H, *Biosens. Bioelectron.*, 2011, **26**, 4169.
12. S. Cheng, B. E. Logan, *Electrochem Commum.*, 2007, **9**, 492.
13. B. E. Logan, S. Cheng, V. Watson and G. Estadt, *Environ. Sci. Technol.*, 2007, **41** 3341.
14. Y. Qiao, S. J. Bao and C. M. Li, *Energy Environ. Sci.*, 2010, **3**, 544.
15. B. E. Logan, *New York: Wiley.*, 2008.
16. H. Y. Tsai, C. C. Wu and C. Y. Lee, *J. Power Sources.*, 2009, **194**, 199.
17. S. D. Huang, Y. S. Liu, W. Li and Y. Yin, *Mechanical & Electrical Engineering Magazine.*, 2008, **25**, 35.
18. S. K. Chaudhuri, D. R. Lovley, *Nat. Biotechnol.*, 2003, **21**, 1229.
19. X. Huang, M. Z. Fan, P Liang and X. X. Cao, *China Water & Wastewater.*, 2007, **23**, 8.
20. F. X. Li, Q. X. Zhou and B.K. Li, *Journal of basic science and engineering.*, 2010, **18**, 877.
21. W. F. Kong, Q. J. Guo, X. Y. Wang and X. H. Yue, *Ind. Eng. Chem. Res.*, 2011, **50**,12225.
22. C. Martínez-Huitle, S. Ferro, *Chem. Soc. Rev.*, 2006, **35**, 1324.
23. C. C. Zheng, Q. J. Guo, X. Y. Wang and W. F. Kong, *CIESC J.*, 2012, **63**,1599.
24. P. Liang, X. Huang, M. Z Fan, X. X. Cao and C. Wang, *Appl. Microbiol. Biotechnol.*, 2007, **77**, 551.

COLLECTION OF NANO-TIO$_2$ AEROSOL BY USING A NOVEL WET SAMPLER

P. Mao[1,2], S.Y. Feng[1], Y. Yang[1,2*], S.W. Chen[1], Z.P. Wang[1,2]

[1] School of Environmental and Biological Engineering, Nanjing University of Science and Technology, Nanjing, China
[2] Lianyungang Institute of Nanjing University of Science and Technology, Lianyungang, China

1. INTRODUCTION

In recent years, a growing number of nanometric technologies are exponentially used in current industries, and most of them involve, in many cases, the handling of airborne nanoparticles. Effects related to nanoparticles (<100 nm) large surface-to-mass ratio and quantum effects give them new characteristics compared to materials of the same chemical composition but of larger dimension. And significant production of free radicals can occur on the surface of the nanoparticles.[1-3] Thus, nanoparticles may possess massive potential risks to human health. Furthermore, nanoparticles are more likely to escape pulmonary clearance via macrophages phagocytosis and to interact directly with biological material compared to larger micron-sized particles. Many scientific and government reports indicate that nanoparticles might represent new risks to human health and the environment.[4-6]

Biological responses to nanoparticles might depend not solely on the dimension of the nanoparticles agglomerates but also on the initial size of the agglomerates, and the initial size is a factor that determines their deposition in the lung, their ability to cross biological barriers, and their capacity to reach and enter cellular targets.[7-11] Evaluating health effect to nano aerosols requires their collection and eventual placement onto a water-based medium in preparation for biological analyses. Therefore, the nanoparticles, collected by currently sampling method, such as filtration and impaction, are not suited for further biological analysis.

Many water-based wet samplers have been studied and have shown potential for collecting nano particle. Counterflow jets of droplets collide, and within this process, the aerosol particles are captured into dispersed liquid, and the experimental result prove the quantitative collection of aerosol particles down to 0.3µm (collection efficiency of 100% for DEHS aerosol) in diameter.[12] Ambient particles mixed with saturated water vapor to produce droplets are collected by inertial techniques, and over 80% of all particles (sulfate aerosol) larger than approximately 50nm were collected.[13-15] The nanoparticles, suspended in the industries, are consisted of different particle types (soluble or insoluble particles).

Efficiency studies of above two collection methods have been limited to solubility nanoparticles. Thus, there is a difficult for them to collect nanoparticles in occupations workplaces. However, liquid impingement have been used for collecting insoluble nanoparticles (TiO_2, SiO_2 and polystyrene nanoparticles), but the minimum collection efficiencies are only 3-10% at sizes between 40 and 47nm.[16]

The aim of this study is to design and fabrication of a wet sampler for collection of insoluble nanoparticles directly into liquid media to make nanoparticle suspensions with high efficiency. The sampler combined the theory of ejector produce negative pressure to inhale air and turbulence droplets impact nanoparticles. It mainly has the following three advantages in comparison with the traditional samplers: effectively collect the semi-volatile aerosols; the high collection efficiency; the simple structure and low cost. In here, nanoTiO$_2$ dust, have widespread use in industries, are choose to represent insoluble nanoparticles. Then the optimum parameters (air flow rate, spray flow rate, and sampling time) are found. And the suspension running conditions are studied by Fluent software simulation.

2. EXPERIMENTAL SECTIONS

2.1 Wet Sampler

The self-made wet sampler, is schematically shown in Figure 1, consists of a spray tower, an ejector, a fluid reservoir, two spray nozzles and two membrane pumps. A cylindrical ejector, of which the inlet size is bigger than the outlet, is installed horizontally at the bottom of the fluid reservoir. According to the theory of the Bernoulli equation, with the special structure of the ejector, negative pressure is formed at the lateral inlet by high speed fluid flow which is originated in membrane pump. As a consequence, nano aerosol is aspirated into the spray chamber through sampling orifice, and the mixtures of aerosol and liquid are drawn into the ejector with a high flow rate from spray chamber. Spray nozzles are installed on the top of spray chamber and fluid reservoir to generate liquid droplets with mean diameter of 40μm in the space. Atomization is the process in which a certain volume of liquid is broken into many small drops generating a much increased surface area. In addition, the liquid film, formed by fallen droplets on the guide plate, is used to capture aerosol particles again. As approaching to the fluid reservoir, a majority of aerosol particles are washed out and collected into the liquid. However, those particles still suspended in the air are blended violently with the droplets sprayed out of the spray nozzle at high speed. Two square crossing plates generate a limited space to promote the collection process effectively. A mist eliminator is installed in the outlet tube to baffle large droplets generated by the bubbling process. Ultimately, the purified air is released out of the sampler.

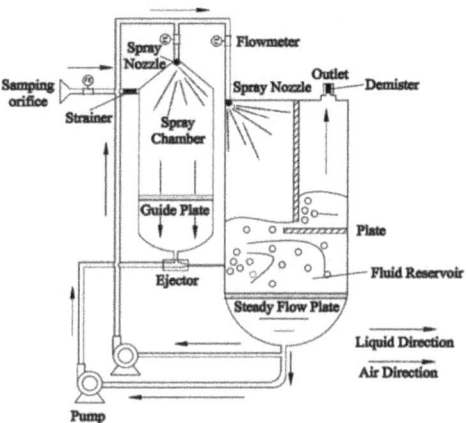

Figure 1 *Schematic of the self-made wet sampler*

2.2 Analytical System

Performance of the wet sampler was tested under laboratory conditions with representative nano hydrophilic aerosol. The hydrophilic nano titanium dioxide was used as representatives in here. Nano TiO_2 aerosol with a standard size distribution of 65±10nm was aerosolized from nano dust by a compressed air with an ejection gun of not less than 4bar pressure. During all experiments, nano aerosol was suspending in a smog chamber, where the temperature and the relative humidity were monitored: T=18-25°C, RH=30-40%. The air and fluid flow through silicone tubing with i.d. 6mm or i.d. 8mm to minimize transport loss in sampling.

Two wet samplers were installed in series to easily calculate its collection efficiency. Aerosol was drawn from the smog chamber and pass through the two samplers sequentially with second distilled water as the collection fluid. Then the ultraviolet wave absorptions of the suspension fluids of the two samplers were measured by UV Spectrophotometer (753N JINGHUA Inc.), and the mass concentrations could be calculated from UV absorptions through correlation curve.

2.3 Collection Efficiency

Prior to determine the collection efficiency of the wet sampler, many ideal conditions were assumed: (1) nano aerosol concentration in smog chamber remained constant; (2) no leakage occurs in sampling system; (3) all parameters of two samplers were same; (4) the loss of nano aerosols in the sampler was zero.

Thus three mathematical model based on mass balance can be obtained.

$$m_1 = Q_G \cdot C_0 \cdot \eta \cdot \Delta \tag{1}$$

$$m_2 = Q_G \cdot C_1 \cdot \eta \cdot \Delta t \tag{2}$$

$$C_1 \cdot V = C_0 \cdot V - m_1 \tag{3}$$

where m_1 and m_2 are the mass of collected aerosol particles in No.1 and No.2 sampler, C_0

and C_1 are the mass concentration of nano aerosol in smog chamber and the outlet of No.1 sampler, Q_G is gas flow rate, η is the collection efficiency of the wet sampler, and Δt is the sampling time.

Integrating Equation (1), (2) and (3) gives

$$\frac{m_2}{m_1} = \frac{V \cdot C_1 \cdot \eta}{V \cdot C_0 \cdot \eta} \tag{4}$$

Hence, the collection efficiency of the wet sampler is calculated as:[17]

$$\eta = 1 - \frac{m_2}{m_1} = 1 - \frac{c''}{c'} \tag{5}$$

where c' and c'' are the mass concentrations of suspensions in NO.1 and No.2 sampler, respectively.

2.4 Collision Mechanism

The sampling approach within this wet sampler mainly consists of droplets collision, liquid film capture, jet impingement, bubbles breaking up, and liquid seal collection, as demonstrated in Figure 1.

Droplets collision probability, based on the O'Rourke model,[18] can be calculated by the following equation.

$$p = \frac{\pi(r_1+r_2)^2 |\bar{v}_r| \Delta t}{V} \tag{6}$$

where p is collision probability, v_r is the relative velocity of two droplets, V is the given volume, r_1 and r_2 are radius of two droplets, Δt is the time step.

Thus, if the particles within a given volume are n_s, the mathematical expectation of collision probability can be calculated as:

$$v = \frac{n_s \pi(r_1+r_2)^2 |\bar{v}_r| \Delta t}{V} \tag{7}$$

From Equation (6) and (7), Particles collision rates in spray chamber can be obtained. Moreover, Qian and Law[14] studied the final outcome of droplets collision, and classified the outcome into five different regimes, named: (I) slow coalescence; (II) bouncing; (III) coalescence after substantial deformation; (IV) reflexive or near head-on separation, and (V) stretching or off-center separation.

Collision outcomes can be determined by comparing b and b_{cr}. If $b > bcr$, outcome is stretching separation; otherwise, outcome is coalescence. In here, impact parameter b, which defines the geometrical orientation of the interacting particles, is orthogonal distance. b_{cr} is obtained by Equation (8).

$$b_{cr} = (r_l + r_s)\left(\frac{min\{1.0, 2.4f(\gamma)\}}{We}\right)^{\frac{1}{2}} \tag{8}$$

where r_l and r_s are radius of large particle and small particle, $f(\gamma)= \gamma3-2.4\gamma2+2.7\gamma$, and $\gamma=r_l/r_s$, and *We* is weber number.

When the droplet impacts the surface, the Sommerfeld parameter (K) characterizes the droplet behavior, when $K<3$, rebound; $3<K<57.7$, deposition; $K>57.7$, splashing. K is calculated as:[19]

$$k = \sqrt{W_e\sqrt{R_e}} \tag{9}$$

where R_e is Reynolds number.

3. RESULTS AND DISCUSSION

3.1 Establishment of Standard Curve

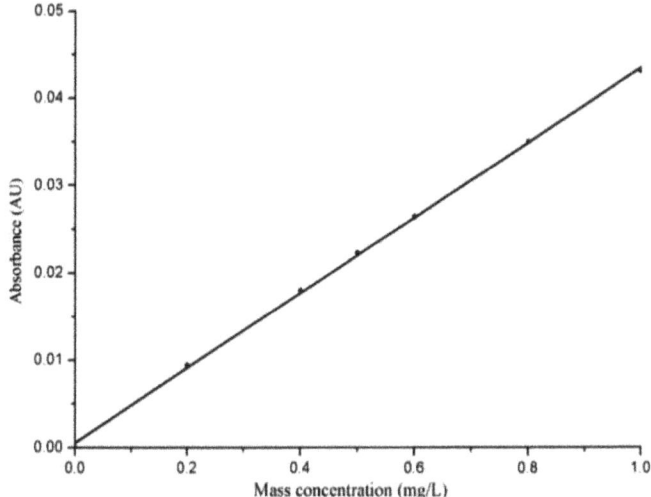

Figure 2 *The ultraviolet absorbance as a function of nano TiO₂ concentrations at 330nm*

The concentrations of the suspensions were analyzed using an ultraviolet spectrophotometer. Prior to measurement, it is essential to plot a spectrophotometer standard curve. Dilute 0.0100 g of the nano TiO₂ dust to 100 ml with redistilled water. One liter of the wording standard is equivalent to 100 mg nano TiO₂. Add 0 ml, 0.5 ml, 1.0 ml, 1.25 ml, 1.5 ml, 2.0 ml and 2.5 ml of the nano TiO₂ working standard solution (1 L=100 mg TiO₂) to a series of seven 250 ml volumetric flasks. To each flask, dilute to the mark with redistilled water. The mass concentrations of resulting solutions are 0.0 mg/L, 0.2 mg/L, 0.4 mg/L, 0.5 mg/L, 0.6 mg/L, 0.8 mg/L and 1.0 mg/L, respectively. Measure the absorbance by ultraviolet spectrophotometry at 330nm. Prepare a standard curve plotting absorbance vs. mass concentration of nano TiO₂, a good linear relationship is shown in Figure 2 and the linear equation is built as: $y = 0.043x + 0.0006$, $R^2=0.9999$.

3.2 Optimization and Efficiency of Sampler

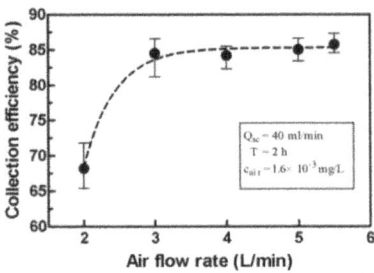

Figure 3 *Effect of air flow rate on the collection efficiency*

The collection efficiency of aerosol as a function of the air flow rate (Q_{air}) in the range 2-5.5 L/min was studied by sampling of nano TiO_2 aerosol followed by measurement of mass concentrations of the suspensions in the two samplers by means of the ultraviolet spectrophotometry. Figure 3 presents the effects of Q_{air} on the collection efficiency at constant aerosol concentration (c_{air}, 1.6×10^{-3} mg/L), spray flow rate in spray chamber (Q_{sc}, 40 ml/min) and sampling time (t, 2 h). It is evident that as the air flow rate increases, the collection efficiency of the wet sampler rapidly increases from 67.6 % at Q_{air} of 2 L/min to 86.8% at 5.5 L/min.

According to Equation (6), collision probability is enhanced with the relative velocity by improving the air flow rates, thus the collection efficiency improved. However, review Equation (7), it is easily obtained that air flow with higher speed will decreasing the collision time and nano aerosol particles in spray chamber, the process of increasing air flow rate is a good servant but a bad master. So too high of air flow rate, i.e. large than 3L/min, the increasing rate of collection efficiency will slow down gradually, as shown in Figure 3.

Figure 4 shows the effect of spray flow rates on the collection efficiency at Q_{air} of 5.5 L/min, c_{air} of 2.2×10^{-3} mg/L and constant sampling time (2h). As shown in Figure 1, two spray nozzles were controlled by one pump. In here, in order to ensure accurate spray flux, sum of two spray flow rates was constant at 80ml/min as shown in Figure 4. The collection efficiency increases from 76.1% at 0ml/min (spray flow rate in spray chamber, Q_{sc}) to a maximum of 92.9% at 40ml/min. However, the collection efficiency decrease along with flow rate increases to 80ml/min.

On the basis of Equation (6) and Equation (7), it isn't difficult to understand the collection efficiency increased with the increasing of spray flow rate. However, collection efficiency declined after spray flow rate increased to 40ml/min. This abnormal phenomenon indicated that function of spray in reservoir play a positive role in nano aerosol sampling, and it is an indispensable part. Comparing the collection efficiency of spray flow rate in the fluid reservoir (Q_{sr}) with Q_{sc} at 80ml/min, it indicates that the effect of the spray nozzle of spray tower more significant on the collection efficiency than the one of the fluid reservoir.

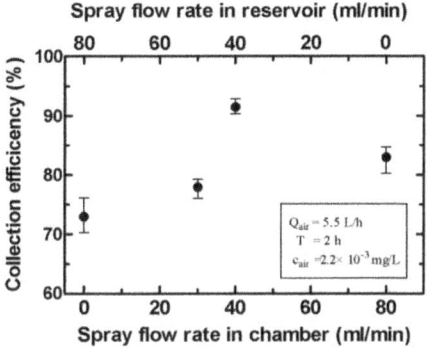

Figure 4 *Effect of spray flow rate on collection efficiency*

Put all above relevant parameters into the equation (8), the result of $b_{cr} > b$ can be obtained, and it is not affected by changes of air flow rate. Thus the outcome of collisions between droplets and particles in spray chamber is coalescence and constant. The outcome of coalescence is helpful for the following wet sampling. And put relevant parameters into the equation (9) too, the results are rebound and deposition. Rebound droplets can collide with particles again, and then collision probability will increase. However, the liquid film, formed by deposition of the droplets, will wash out the particles which stick in the surface of spray chamber.

An interesting phenomenon can be found that the collection efficiency increased from 86.8% to 92.9% with the increasing of the mass concentration of aerosol, if comparing Figure 3 with Figure 4 at the same other parameters. It same can be explained according to Equation (7), particles increase with the increasing of aerosol concentration.

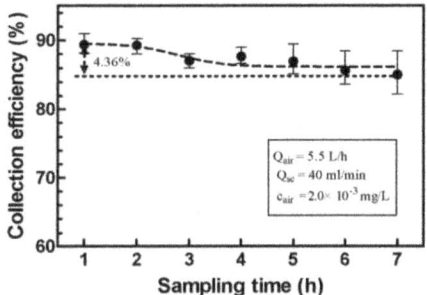

Figure 5 *Effect of sampling time on collection efficiency*

The relationship of the sampling time between the collection efficiency is investigated at the optimum parameters of the sampler, at Q_{air} of 5.5L/min and Q_{sc} of 40ml/min. As shown in Figure 5, although collection efficiency of the sampler is decreasing with the sampling time on the whole, the decreasing gap is only a few percent, from 89.37% to 85.01%, in 7 hours of sampling. The primary cause for decline in collection efficiency could be increased mass concentration of suspension. However, the weak dependence of collection efficiency on sampling time demonstrates a high performance of the sampler.

3.3 Fluent Software Simulation

The goal here is to use Fluent software as a tool to master the movement of suspension in reservoir. Figure 6 and Figure 7 show the simulation results. It indicates that suspension mainly swirls in a clockwise motion in the fluid reservoir. Eddy motion formed mainly because of suspension impact to surface. And suspension movement is affected by steady flow plate. Many small vortex motions are existence under the steady flow plate and the sides of reservoir. These eddy motions will produce effective collection functions of inertial impaction and impingement.

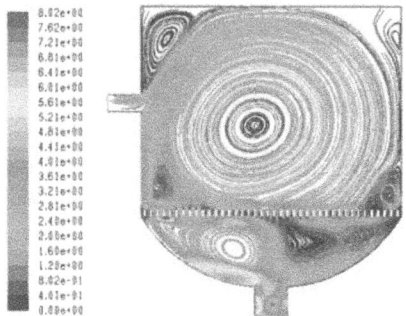

Figure 6 *Pathlines shaded by velocity magnitude (m/s)*

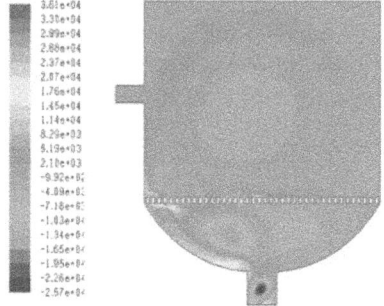

Figure 7 *Contours of total pressure (Pa)*

4. CONCLUSIONS

A novel wet sampler for sampling of nanometer aerosol particles is developed. It is found that the sampler's collection efficiency increases with increasing the air flow rate and fluid flow rate, but when spray flow rate higher than 40ml/min, the collection efficiency is decreased. Weak dependence of the collection efficiency was found on the sampling time. An optimum collection efficiency of 92.9% for nano TiO_2 aerosol has achieved by sampling at air flow rate of 5.5L/min and spray flow rate of 40ml/min for 2 hours. Furthermore, Fluent software simulation showed that suspension clockwise vortex movement in the reservoir. The simple structure and high collection efficiency performance makes the sampler suitable for the sampling of many kinds of nanoparticles.

ACKNOWLEDGMENTS

This work was supported by the Foundation of Jiangsu Environmental Protection Department (No. 2012015), Lianyungang Social Development Projects (No.SH1203), Lianyungang Infrastructure Project (No.JC1203) and Financial Supported by Lianyungang Institute of NJUST, China (No.NUSTLYG-2012-001)

References

1   K. Donaldson, P. Beswick, P. Gilmour. *Toxicology Letters*, 1996, **88**, 293.
2   C.A. Dick, D.M. Brown. *Inhalation Toxicology*, 2003, **15**, 39.
3   B. Hervé-Bazin. *Les nanoparticules: Un enjeu majeur pour la santé au travail?* EDP Science, France, 2007, p. 701.
4   National Institute for Occupational Safety and Health (NIOSH). *Occupational Exposure to Titanium Dioxide*. DHHS (NIOSH) Publication, USA, 2011, p. 1.
5   J.J. Li, S. Muralikrishnan, C.T. Ng, L.Y. Yung, B.H. Bay. *Experimental Biology and Medicine*, 2010, **235**, 1025.
6   D.B. Warheit. *Analytical and Bioanalytical Chemistry*, 2010, 398, 607.
7   C. Foged, B. Brodin, S. Frokjaer, A. Sundblad. *International Journal of Pharmaceutics*, 2005, **298**, 315.
8   A. Nemmar, M.F. Hoylaerts, P.H. Hoet, J. Vermylen, B. Nemery. *Toxicology and Applied Pharmacology*, 2003, **186**, 38.
9   G. Oberdörster, E. Oberdörster, J. Oberdörster. *Environmental Health Perspectives*, 2005a. **113**, 823.
10 E. Koike, T. Kobayashi. *Chemosphere* 2006, **65**, 946.
11 K. Donaldson, A. Seaton. *Journal of Nanoscience and Nanotechnology*, 2007, **7**, 4607.
12 P. Mikuaka and Z. Ve eǎ. *Anal. Chem.* 2005, **77**, 5534.
13 A. Khlystov, G.P. Wyers, J. Slanina. *Atmospheric Environment*, 1995, **29**, 2229.
14 J. Qian, C.K. Law,,. *J. Fluid Mech.*. 1997, **331**, 59.
15 R.J. Weber, D. Orsini, Y. Daun, Y.N. Lee, P. Klotz, F. Brechtel. *Aerosol Science Technology*, 2001, **35**, 718.
16 Z.C. Wei, R.C. Rosario, L. D. Montoya. *Atmospheric Environment*, 2010, **44**, 872.
17 Y. Yang, S.W. Chen, Z.P. Wang, et al. *The assessment method of collection efficiency in micro-nano dust sampler*. China Patent, 2010, CN 101694428 A.
18 P. J. O'Rourke. *Collective drop effects on vaporizing liquid sprays*. Ph.D. thesis, Princeton University, USA, 1981.
19 C. Escure, M. Vardelle, P. Fauchais. *Plasma Chemistry and Plasma Processing*, 2003, **23**, 185.

INVESTIGATION OF DROPLET COALESCENCE AND OIL-WATER SEPARATION CHARACTERISTICS OF INSULATED ELECTRODE IN ELECTRIC DEHYDRATOR

Y.L. LV[1*], Q. ZHANG[2], L.M. HE[1], X.M. LUO[1]

[1] Department of Storage & Transportation Engineering, China University of Petroleum, NO.66 Changjiang West Road, Qingdao, China
[2] Department of Technical Equipment, Dwell Company Limited, No.8 Xinxi Road, Shangdi, Haidian District, Beijing, China

1   INTRODUCTION

The destabilization of water-in-oil emulsions is an important industrial process,[1] electrostatic coalescence is the most efficient demulsifying technology having been adopted in oilfield operations .[2] In China, traditional electric dehydrators with bared electrode breakdown frequently. As the oil field exploitation entering the middle and later periods, the oil-water separation in the electric dehydrator will be more and more difficult because of the use of the polymer flooding and varieties of additives. [3,4]

For the electric dehydration process, the modification of internal electrode structures and the optimization of operating parameters are imperative. And insulated electrode can be used to solve these problems. The droplet coalescence properties of the lactescence between insulated electrodes in the new dehydrator have been studied systematically. The study on oil-water separation characteristics of insulated electrode provides some significant conclusions for engineering practice.

2   METHOD AND RESULTS

2.1 The Experimental System

Transparent electric dehydrator and flat insulated electrodes were designed to study the droplet coalescence and oil-water separation characteristics of insulated electrode under various electrical and flow parameters. The experimental system was shown as Figure 1. Several different oil emulsions (two kinds of white oils and two kinds of crude oils) were prepared, the isokinetic sampling device and the photomicrography system were used to gather droplets data. The processes of droplet coalescence and oil-water separation in transparent dehydrator were recorded by video camera simultaneously. Based on the small-

Figure 1 *The experimental system*

sized electric dehydrator, the droplet coalescence properties of different electric field parameters, different flow field parameters and different water-oil emulsions have also been discussed.

2.2 Result and Discussion

2.2.1 Electric Field Parameters.
The study results showed that the oil-water separation efficiency was closely related to the electric field intensity, frequency, oil property, component, flow rate and water content.

Electric field intensity was the key factor while frequency had a bit impact on separation efficiency. The changes of droplets flowed through electrode with and without voltage were shown as Figure 2.It was found that along with the electric field intensity increased, the droplet coalescence phenomenon became more obvious, the coalescence response time diminished and the oil-water separation efficiency increased strikingly(Figure 3). But when the electric field intensity surpassed the critical surface tension, large droplet would break up into many small droplets, which was also called electric dispersion phenomenon(Figure 4).

The experimental results showed that crude oil emulsion droplets changed sensitively with frequency, while white oil droplets had no obvious change(Figure 5).

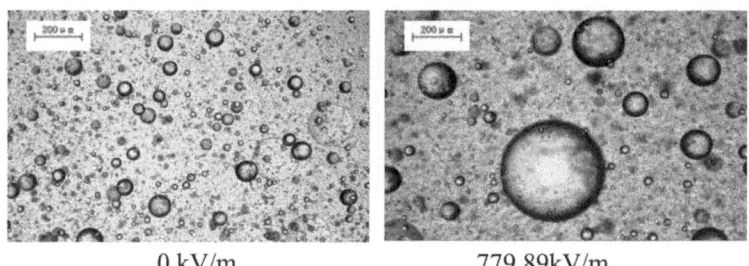

0 kV/m   779.89kV/m

Figure 2 *The droplet changes at different electric field intensity*

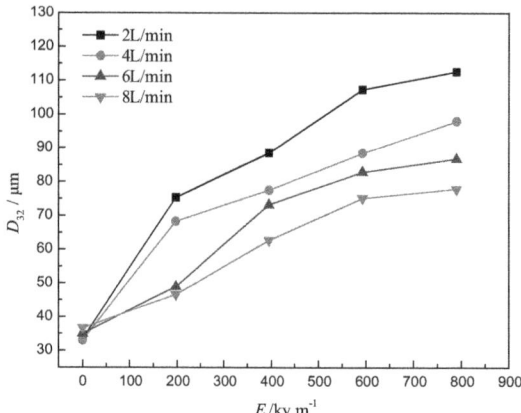

Figure 3 *Flow rate vs. E-field strength in 10% water content*

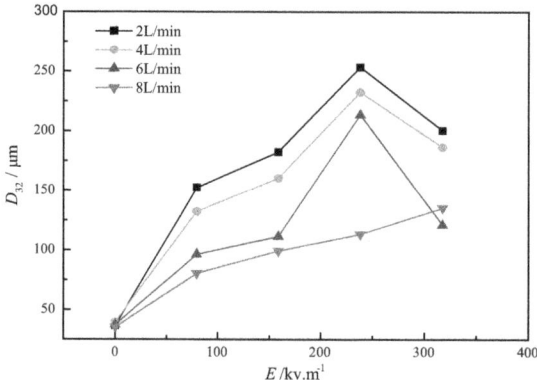

Figure 4 *Flow rate vs. E-field strength in 30% water content*

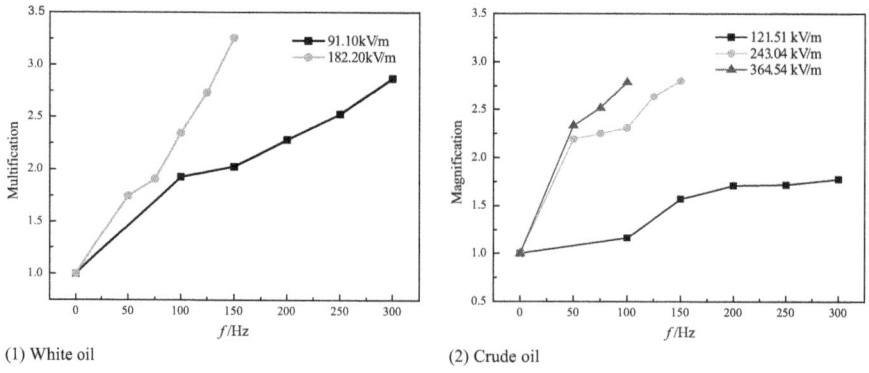

Figure 5 D_{32} *vs. frequency at different electric field intensity*

2.2.2 Operation Parameters.

The change of operation parameters would directly influence the droplets coalescence process and oil-water separation efficiency of the electric dehydrator. The influence of flow rate on separation is complex (Figure 6). When high viscosity oils were investigated, a high water content emulsion layer would easily form at a low flow rate (Figure 7). Because of this layer, the separation efficiency decreased, the electric field intensity was weakened, the scouring force exerted on the layer reduced with the decrease of flow rate, and eventually the oil-water separation were influenced.

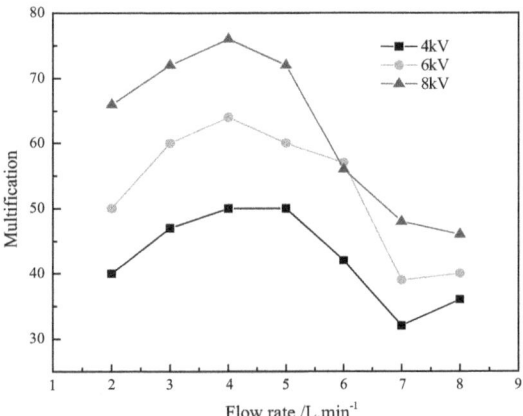

Figure 6 *The droplet magnification vs. flow rate at different electric field intensity*

As to different oil-water emulsions, the required electrode gap and flow rate were different. With the increase of temperature, the droplet coalescence effect rose (Figure 8). As the temperature increased, the viscosity of the continuous phase decreased, the oil-water interfacial strength reduced and the drain speed of the emulsion layer increased, all of which benefited the droplet coalescence.

Figure 7 *The oil-water separation phenomenon*

2.2.3 Oil Properties.

When the viscosity of the continuous phase was lower, the oil component was simpler; the electric dehydrator efficiency went up (Figure 9). Because when the viscosity of the continuous phase was lower, the droplet collision probability would be enlarged.

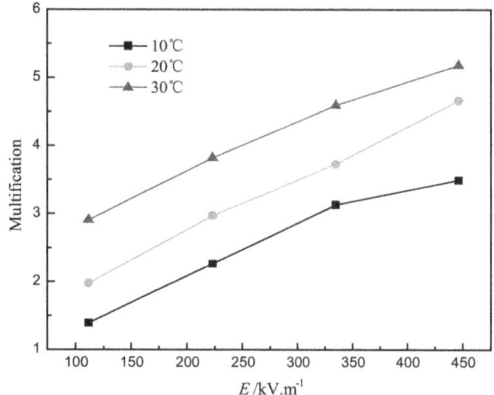

Figure 8 *The droplet magnification vs. electric field intensity at different temperature*

Under the same electrical condition, the separation efficiency of crude oil emulsion was higher than white oil (Figure 10). That's because the electrical conductivity of crude oil was higher than the latter, the electric filed force operated on droplets was lower, which led to difficulty of coalescence. Although the viscosity and electrical property of the two kinds of crude oil were close, the droplet coalescence laws of the two oils differed a lot, for the complex oil components and the contents of the natural emulsifying agents made each kind of oil unique.

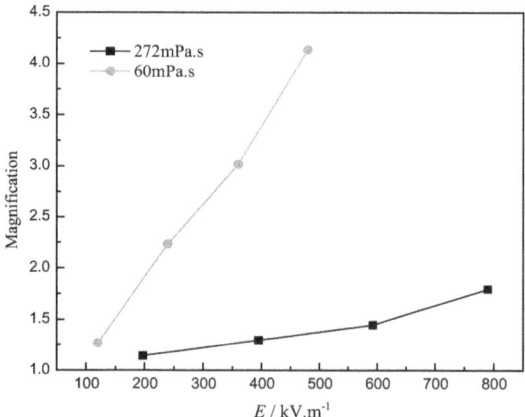

Figure 9 *The droplet magnification vs. E-field strength at different viscosity*

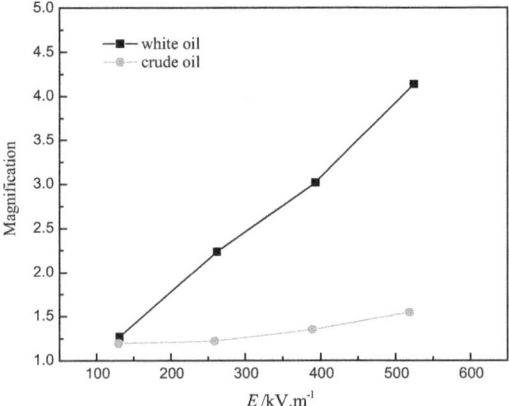

Figure 10 *The droplet magnification vs. E-field strength at different oil properties*

When oil was the continuous phase, the droplet coalescence phenomenon became more obvious with the increase of water content. What's more, the oil-water separation turned into more effective with the rise of droplet size. The droplet diameter of dispersed phase was the key factor that influenced the oil-water separation efficiency. If the droplet diameter was too small, the droplet coalescence phenomenon would hardly occur. Experimental results showed that AC electric field was more suitable for dealing with high water content emulsions. The higher water content was, the higher permittivity of emulsion was, and the liquid film broke up much easier, which was helpful to droplet coalescence.

3   CONCLUSION

In this work we have considered droplets coalescence properties and W/O emulsion separation characteristics. Several diagrams have been plotted and the effect of each parameter on the coalescence has been investigated.

According to the results, the electric field intensity was the key factor while frequency had a bit impact on separation efficiency. The electric dispersion phenomenon occurred at high electric intensity. A high water content emulsion layer would easily form at a low flow rate. When the viscosity of the continuous phase was lower, the electric conductivity was smaller; the electric dehydrator efficiency went up. On the other hand, complex oil components and the contents of the natural emulsifying agents made each kind of oil unique. Since the affecting factors are very complicated, performing further electrode design and laboratory experiments will result in better results for industrial applications.

ACKNOWLEDGEMENTS

The work is partially supported by a grant from Chinese National Natural Science Foundation (Grant No. 51106182, 51274233, 51006124). Natural Science Foundation of Shandong Province of China (ZR2010EQ040).

References

1. I.G. Harpur, N.J. Wayth, A.G. Bailey., *J. Electrostatics.*, 1997, **40**, 135.
2. C. Noik, J. Chen and C. Dalmazzone, SPE103808.
3. J.S. Eow, M. Ghadiri, *Chem. Eng.J.*, 2002, **85**, 357.
4. I. B. Ivanov, D. S. Dimitrov and P. Somasundaran, *Chem. Eng. Sci.*, 1985, **40**, 137.

COALESCENCE AND MOVING CHARACTERISTICS OF DROPLETS UNDER PULSED DC ELECTRIC FIELD

D.H. Yang, M.H. Xu, L.M. He*, Y.L. Lü, H.P. Yan
College of pipeline & Civial Engineering, China University of Petroleum, No.66 of Changjiang west road, Qingdao, China

1 INTRODUCTION

Dehydration of water in oil (W/O) emulsion becomes difficult as the crude oil properties change continuously with oil production, especially when a large amount of polymer is added to enhance oil production. Separation methods utilizing high electric fields have been used extensively in petroleum industry.[1] Most of the conventional electro-separators are huge and bulky, thus having high capital and operating costs. A compact electrocoalescer will have the advantage of being smaller and lighter than traditional ones and thus easier to install on an offshore platform and reduce the investment needed. There is, therefore, great scope to optimize the design and operation of these separators based on a fundamental understanding of parameters that affect the coalescence of aqueous drops in oil. However, the fundamental mechanisms governing the behavior of water in oil emulsion under a high electric field are still not very well understood, due to the complex interaction among various influencing factors.

Researchers have paid attention to droplet deformation and coalescence for a long time. With the development of low-cost high speed cameras, studying the deformation and movement of micro droplets is now much more practical. Eow[1] investigated the deformation of aqueous drops in several liquid/liquid systems under a DC electric field with a high speed camera. The diameter of the drops was about 1 mm. The effect of electric field intensity and interfacial tension were analyzed. Before rupture, the droplet is elongated and the deformation degree is proportional to the electric field intensity. The degree of deformation is different in different liquid/liquid systems, depending on the physical properties of the respective systems. Eow[2,3] also found the shape of drops is related to conductivity, viscosity of the continuous phase, interfacial tension and viscosity of the drop liquid. The deformation degree is small under a small dielectric constant of oil and large interfacial tension. Supeene[4] explored the influence of the fluid, interfacial, and electrical properties on the system dynamics by a numerical method. The simulations were compared with analytic solutions, as well as available experimental results, which indicated that the predictions from the model were reliable even at considerably large deformations.

The relative movement of two droplets in an electric field is governed by the forces acting on the droplets. Bails [5] and Taylor [6] indicated that field-induced charges on the water droplets cause adjacent droplets to attract each other. The electrically induced force was the most important force to promote droplet coalescence. Eow did a lot of work on droplet coalescence. Eow[7,8] reviewed the development of dehydrating technology and the development of investigation on droplet coalescence. Eow[7] pointed out that under a high strength electric field or when the center-to-center distance of two droplets was short, the result calculated with the spherical droplet polarization model was not accurate. Eow[9] investigated the effects of the applied electric field, angle between the electric field and the center line of the two drops, and also electrode geometry to droplet coalescence under a pulsed electric field. Eow[10] also studied droplet coalescence in a new dehydrator. Chiesa[11-13] analyzed all the forces acting on water droplets and simulated their movement process numerically. Also, they investigated the coalescence process of a small droplet falling onto a stationary bigger one experimentally, and compared the experimental data with the numerical simulation results. A lot of papers also focus on the effect of electric field intensity and frequency on the coalescence efficiency of the dehydrator.[14,15]

Bales and Stitt[16] believed that dipole coalescence is the main reason for droplet coalescence in AC and DC electric fields. Bailes and Larkai[17] published a theory that dielectric relaxation was important during droplet coalescence and before coalesce droplets form chains. Chen[18] also found in experiments that droplet chains are short in a high frequency field while long in a low frequency field. Taylor[19] pointed that when an AC electric field is applied very stable droplet chains form between the two electrodes, which can cause current leakage and prevent droplet coalescence.

Many researchers focused on the deformation, moving of one droplet or two droplets, and the droplet diameter is about 1mm. But droplets in W/O emulsion are usually smaller than 100 μm. Seldom investigation is aimed at the coalescence characteristics of droplets less than 100 μm in water/oil emulsion. In this paper, experiments were carried out to investigate the moving of multi droplets in water/oil emulsion under pulsed DC electric field. The moving velocity, moving direction and coalescence manner of droplets less than 100μm under frequencies from 10Hz to 2000Hz and electric field strength of 500kV/m were studied with a micro high-speed camera system and image processing technology.

2   METHOD AND RESULTS

2.1 Experimental facility

The micro-experimental setup used for investigation is shown in Figure 4. The experiment set-up includes high-speed high-voltage power amplifier, high-speed digital camera and experimental cell. The electrodes were fixed on opposite sides of the rectangular Perspex

Figure 1 *High-speed high-voltage power amplifier*

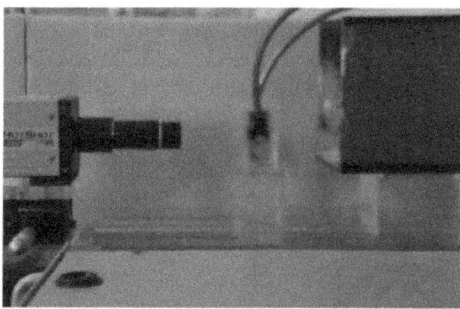

Figure 2 *High-speed digital camera*

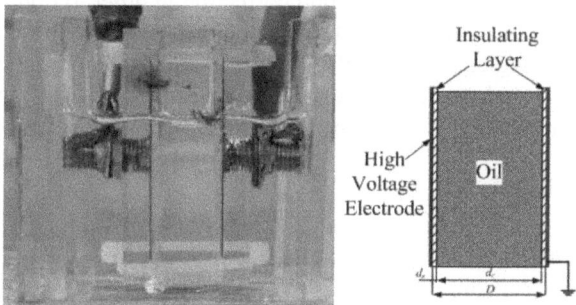

Figure 3 *Experimental cell*

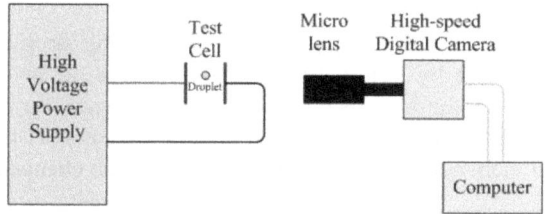

Figure 4 *Schematic diagram of the experimental system*

cell. The high-speed high-voltage power amplifier (Trek 20/20C) is made in America and can supply different wave forms, such as sine AC , DC, pulsed DC and any other forms by entering parameters. The voltage is from -20 to 20 kV and frequencies 1-10000 Hz. The deformation and movement of emulsion drops in the pulsed DC electric field was imaged with a high speed camera (Figure 2) equipped with microscopic lenses and continuously recorded by a computer. The total optical magnification was 400. Images can be processed by image processing technology. The ordinate of droplet center and velocity can be got.

White oil and tap water were chosen as the continuous phase and dispersed phase, respectively. The properties of white oil and distilled water are presented in Table 1. Emulsion is prepared by emulsifier and then injected to the center of the experimental cell with syringe.

Table 1 *Properties of the oil and water at* 20°C

Medium	Viscosity/mPa·s	Relative dielectric constant	Conductivity/S·m⁻¹	Density/kg·m⁻³	Surface tension/mN·m⁻¹
White oil	538	2.22	1×10^{-14}	868.5	31.7
Tap water	1.0087	82	3.23×10^{-2}	998.2	72

2.2 Method for analyzing drop images

The resolution of images from the high speed camera is 1280×1024. Before the experiment, the camera images were calibrated at 2.84 μm/pixel. Image-Pro Plus was used to manipulate the captured images. With this software, images can be processed by filtering, fast Fourier transform (FFT) and background correction in order to optimize the quality. In the image, we can choose measurement ranges and measure object properties, such as distance, area, center, angle, circle length, long axis and short axis. Also we can choose automatic or manual mode to calculate other properties, such as vibration frequency, deformation degree, relative velocity, movement trail and so on.

2.3 Results and discussion

2.3.1 Coalescence Manner.
The results (Figure 5) indicate that the coalescence manner is mainly dipole coalescence and electrophoresis coalescence while droplet chain is not found under pulsed DC electric field. At first droplets coalesce in dipole coalescence manner to form larger droplets, then larger droplets will move to a certain electrode because of electrophoretic. During moving, droplets coalese in two manner. One is droplets with high velocity colliding with ones of smaller velocity, which cause collision coalescence. The other is diplole coalescence for the effect of dipole force.

As shown in Figure 5 (d)~(g), there is vortex during droplets moving for the disturbance of electric field by larger droplets. This will result in irregular moving of droplets, which is benificial for droplets collision and coalescence.

We can also find some other phenomenon (Figure 7). Such as this 3 droplets move together and then 2 of them coalesce and then moving direction changed. This is mainly

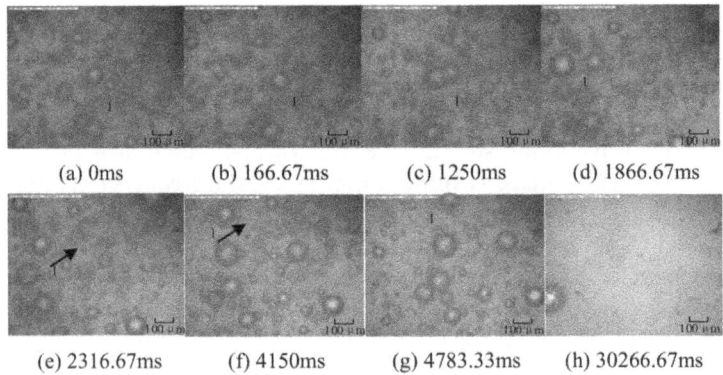

(a) 0ms	(b) 166.67ms	(c) 1250ms	(d) 1866.67ms
(e) 2316.67ms	(f) 4150ms	(g) 4783.33ms	(h) 30266.67ms

Figure 5 *Moving of droplets (f=10Hz, E=500kV·m⁻¹)*

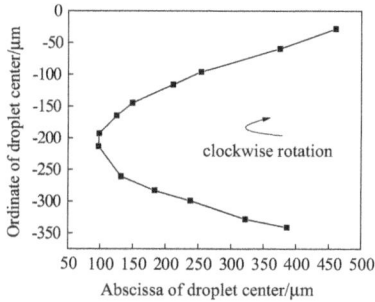

Figure 6 *Trajectories of droplets in regime 5*

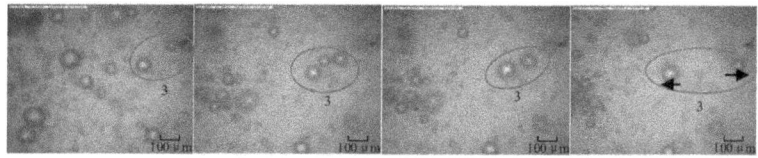

Figure 7 *Changes of droplets (f=20Hz, E=500kV·m⁻¹)*

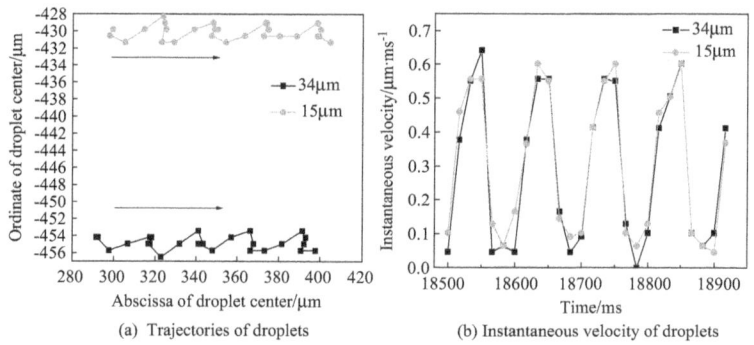

Figure 8 *Trajectories and velocity of two drops (f=10Hz, E=500kV·m⁻¹)*

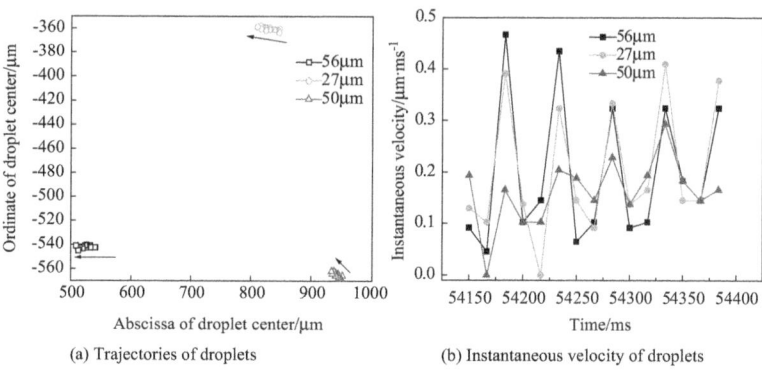

Figure 9 *Trajectories and velocity of two drops (f=20Hz, E=500kV·m⁻¹)*

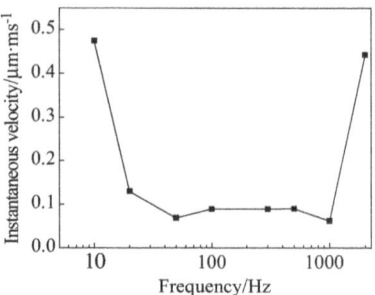

Figure 10 *Average velocity of droplets under different frequencies*

for the charge of different drops is different, so after coalescence combination of charge may change charge polarity and the force is opposite, as a result moving direction change.

If there are seldom droplets the moving direction is in line with the direction of electric field for the electric field strength is almost the same.

2.3.2 Influence of Frequency.

During electrophoresis, droplet velocity is related to the frequency of the electric field. When frequency is 10Hz, droplets moved to the right electrode under the effect of electric field. As shown in Figure 8 (a), the droplet moved in direction composed of vertical downward and horizontal right. The droplet moved downward for gravity, and also moved to right, in the direction of maximum field strength.

But the main moving direction is horizontal right for electric force is much higher than gravity. From Figure 8 (b) it can also be found droplet velocity varies periodically with a period of 100ms, which is consistent with the period of the electric field.

When the frequency is 20 Hz (Figure 9), the period is 50 ms, being consistent with the period of electric field. Droplets did not move in the same direction for droplets are dispersed in different region, so the electric field intensity and the dipole electric field force are not the same, resulting in different moving direction.

When frequency is 10Hz (Figure 8(b)), the minimum velocity of different droplet is all close to 0m/s. In a period of pulsed DC electric field, when power is off, the droplet moving is prohibited by the viscous resistance from continuous phase and the velocity will soon reduce to 0m/s. However, when power is on the droplet velocity will soon rise, meaning that droplet can effectively respond to the change of electric field under low frequency. While frequency is 20Hz (Figure 9(b)) the minimum droplet velocity will be gradually higher than 0m/s, which indicates that when the frequency increases, droplet can not timely respond to the change of electric field. That is to say, during the time of power off, droplet velocity decreases but is higher than 0m/s. And then the electric field is on, so the droplet velocity will also rise. But because of shorter period and shorter acceleration time, the maximum droplet velocity will decrease.

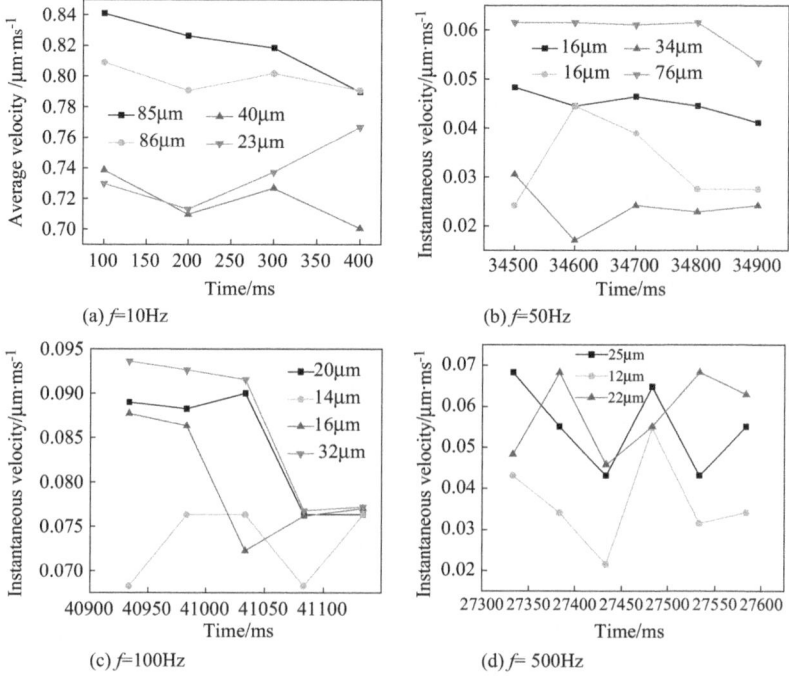

Figure 11 *Velocity of droplets under different frequencies when E=500kV·m⁻¹*

The average velocity of droplet with diameter of 50μm under different frequencies is shown in Figure 10. It can be found that starting from 10Hz, the droplet velocity decreases with the increase of frequency. After the frequency increases to 50Hz, velocity is almost the same from 50 to 1000Hz. But when the frequency increased to 2000Hz, the velocity accelerates. This means there are two best frequencies in low and high frequency interval. This can be explained by the relaxation time. According to the Bailes formula (1),[20] the optimal frequency is much less than 10Hz, so velocity is largest under frequency of 10Hz in low frequency interval. Under frequency of 10Hz the droplets have sufficient time to response to the change of electric field and acceleration time is enough, so the velocity is larger. While under high frequency of 2000Hz, the droplet can not timely respond to change of the electric field, resulting in little change of droplet velocity and larger average velocity.

$$f_p = \frac{1}{2\pi\tau_{\text{OFF}}} \tag{1}$$

$$\tau_{\text{OFF}} = \frac{(C_i + C_e)R_i R_e}{R_i + R_e} \tag{2}$$

2.3.2   *Influence of Droplet Diameter.*

Droplet size also have influence on moving velocity and trajectory of droplet. When frequency is 10Hz, the velocity of big droplet is higher than that of small one. This is mainly for the electric field intensity is proportional to the third power of droplet diameter, while drag force is proportional to the square of the droplet diameter, but electric field force is much larger than the drag force. As a result larger droplet leads to greater resultant of forces, bigger acceleration and higher velocity. Under other frequencies, the phenomenon is also the same (Figure 11).

3   CONCLUSION

The results indicate that the coalescence manner is mainly dipole coalescence and electrophoresis coalescence while chain coalescence is not found under pulsed DC electric field. At first droplets coalesce in dipole coalescence manner to form larger droplets, and then move in a certain direction like electrophoresis. There is vortex during the moving process of droplets for the disturbance of electric field by larger droplets. If there are seldom droplets the moving direction is in line with the direction of electric field for the electric field strength is almost the same. The moving velocity is influenced by frequency and there exists two maximum values under frequencies of 10Hz and 2000Hz. The velocity of droplets is periodical and the frequency is in the same with the frequency of electric field. The velocity is also affected by droplet diameter and larger droplets aid to larger velocity, but droplets nearby move almost in the same velocity.

ACKNOWLEDGEMENTS

The work is partially supported by a grant from Chinese National Natural Science Foundation (Grant No. 51106182, 51274233, 51006124). Natural Science Foundation of Shandong Province of China (ZR2010EQ040)

References

1   J. S. Eow, M. Ghadiri and A.Sharif, *Colloids Surf. A: Physicochem. Eng. Asp* , 2003, **225**, 193.
2   J. S. Eow, M. Ghadiri and A.Sharif, *J. Electrostatics.*, 2001, **51-52**, 463.
3   J. S. Eow, and M. Ghadiri, *Chem. Eng. Process.*, 2003, **42**, 259.
4   G. Supeene, C. R. Koch and S. Bhattacharjee, *J. Colloid Interface Sci.*, 2008, **31**, 463.
5   P. J. Bailes and S. K. Larkai, *Trans. Ichem E.*, 1981, **59**, 229.
6   S. E. Taylor, Trans. *Ichem E.*, 1996, **74**, 526.
7   J. S. Eow, and M. Ghadiri, *Chem. Eng. J.*, 2001, **84**, 173.
8   J. S. Eow, and M. Ghadiri, *Chem. Eng. J.* 2002, **85**, 357.
9   J. S. Eow, and M. Ghadiri, *Colloids Surf. A: Physicochem. Eng. Asp.*, 2003, **219**, 253.
10 J. S. Eow, M. Ghadiri and A. Sharif, *J. Pet. Sci. Eng.*, 2007, **55**, 146.
11 M. Chiesa, J. A. Melheim and A.Hannisdal, *J. Disper. Sci. Technol.*, 2005, **26**,615.
12 M. Chiesa, J. A. Melheim and A. Pedersen, *Eur. J. Mech. B Fluids*, 2005, **24**, 717.
13 M. Chiesa, S. Ingebrigtsen and J. A.Melheim, *Sep. Puri. Technol.*, 2006, **50**, 267.
14 Y. Zhang, Y. Liu and R. Ji, *Colloids Surf. A: Physicochem. Eng. Asp.*, 2011, **373**, 130.
15 G. Harpur, N. J. Wayth and A. G. Bailey, *J. Electrostatics.*1997. **40**, 135.
16 P. J. Bailes and E. H. Stitt,. *Chem. Eng. Res. Des.*, 1987, **65**, 514.

17 P. J. Bailes and S. K. L. Larkai, *Trans. Inst. Chem. Eng.*, 1982, **60**, 115.
18 T. Y. Chen, R. A. Mohammed and A. I. Bailey, *Colloids Surf. A: Physicochem. Eng. Asp.*, 1994, **83**, 273.
19 S. E. Taylor, *Colloids Surf.*, 1988, **29**, 29.
20 P.J. Bailes, S.K.L. Larkai, *IChemE Symp. Ser.*, 1984, **94**, 235.

APPLICATION OF ULTRASONIC METHOD ON PARTICLE CONCENTRATION IN GAS-LIQUID TWO-PHASE FLOW

Anli YUAN, Mingxu SU, Yongming LI, Xiaoshu CAI, Pengfei YIN

Institute of Particle & Two-Phase Flow Measurement, University of Shanghai for Science and Technology, Shanghai, 200093, China

1   INTRODUCTION

The gas-liquid two-phase flow is very familiar to our daily life and industries. It involves many fields including electricity, chemistry, food processing and so on. To achieve the goal of optimal control of the process and reduce energy consumption, the measurement of particle size and concentration (or wetness) in gas-liquid two-phase flow is very significant. For example, the wet steam in last stages of low pressure (LP) steam turbines can yield decreased efficiency and turbine blade erosion in large power plants and resultantly affect the efficiency of generating electricity of power plants. It is critical to perform an on-line measurement of wet steam in last stages of LP steam turbines [6].

The traditional methods of measuring concentration of wet steam are mainly: 1) thermodynamic method; 2) optical method; 3) microwave method; 4) capacitance method; 5) electrostatic measurement method, etc. These methods have revealed their advantages and disadvantages in different environments [2]. Of all the methods described above, the optical method is most widely used in the measurement of wet steam now, because it can present a non-destructive and on-line way to extract the information of concentration and droplet sizes simultaneously. However, when the concentration of wet steam is very high, the attenuated optical signal can't be received correctly by optical detectors, then the accuracy of measurement of concentration will be seriously affected. To overcome this difficulty, a low-frequency ultrasonic wave has been chosen as an information carrier of droplets concentration, in that the ultrasonic method possesses the advantages of strong ability to penetrate which can bring forward a solution for the problem of high concentration measurement. It makes sense to measure the concentration of wet steam by ultrasonic method as a new measurement technique. Nowadays the research of this method is rare. This research carried out by ultrasonic method has a very positive meaning.

As an investigation on the fundamental of measurement, in this paper the wet steam was replaced by a two-phase system of water droplets (generated by a humidifier) in the air. After simulating the related experimental environment according to the wet steam measurement of last stage of steam turbine, we could measure the concentration of droplets by ultrasonic method.

2 THEORETICAL PRINCIPLE AND NUMERICAL COMPUTATION

2.1 Ultrasonic attenuation model

In the gas-liquid two-phase flow, the ultrasonic propagation can be described using classical ECAH (Epstein-Carhart-Allegra-Hawley) model, which considers viscous, heat, scattering and absorption effects in the propagation process, and predicts ultrasonic attenuation of ultrasound after passing through the gas-liquid two-phase flow. As a classic model in the measurement of ultrasonic particle sizing, the ECAH model gives complex wave number expression of two-phase system

$$\left(\frac{k}{k_1}\right)^2 = 1 + \frac{3\emptyset}{jk_c^3R^3}\sum_{n=0}^{\infty}(2n+1)A_n \tag{1}$$

By definition,

$$k = \omega/c_s(\omega) + j\alpha_s(\omega) \tag{2}$$

In Eq.(2), α_s and c_s denote the attenuation coefficient and the speed of sound respectively which are associated with the frequency of ultrasound in the two-phase system. Assuming volume concentration $\emptyset$ and incident compression wave number k_1 as known parameters, the prediction of ultrasonic attenuation and velocity depends on the scattering coefficient A_n. Details of the calculation can be found in the reference [7].

2.2 Numerical computation

Physical parameters of droplet and air used in the ECAH model have been listed in Table 1. Some of the physical parameters are very important to our numerical computation, for example the density of droplet. Figure 1 illustrates attenuation values were calculated by both 25kHz and 40kHz ultrasonic frequencies based on the well-known ECAH (Epstein-Carhart-Allegra-Hawley) model by the physical parameters in Table 1.

Table 1 *Physical parameters of droplet and air(20 ℃)*

	ρ /kg·m^{-3}	c /m·s^{-1}	α /Np·m^{-1}	C_p /J·kg^{-1}·k	τ /W·m^{-1}·k	β /K^{-1}	η /Pa·S	μ /N·m^{-2}
droplet	997.0	1496.7	2.2e^{-14}f	4178.5	0.5952	2.57e^{-4}	—	9.03e^{-4}
air	1.29	343	1.7e^{-11}f	1004	0.0259	3.661e^{-3}	1.81e^{-5}	—

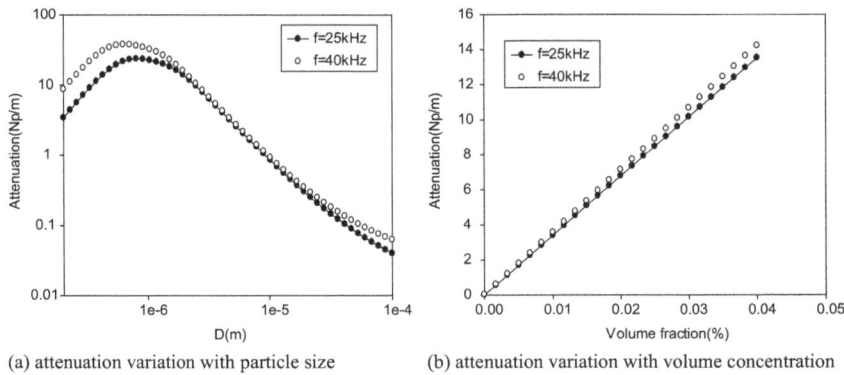

| (a) attenuation variation with particle size | (b) attenuation variation with volume concentration |

Figure 1. *Theoretical ultrasonic attenuation in a droplet/air two-phase system*

The relationship between ultrasonic attenuation and the particle size and volume concentration of droplets at two frequencies of 25 kHz and 40 kHz was shown in Figure 1. As showed in Figure 1(a), the ultrasonic attenuation is sensitive to the variation of particle size when the size was between 1~10μm and particle size of droplets can cause a great effect on the ultrasonic attenuation.

In Figure 1(b), the volume concentration, ranging from 0.0017% to 0.04%, is very low by comparing with conventional ultrasonic measuring conditions. Thus the multiple scattering effect and particle-particle interaction effect can be eliminated safely, and ultrasonic attenuation increases linearly with the concentration.

When the measured environment is made up of single-dispersed particles, the dual-frequency method can be used to get the concentration as a useful method to realize a fast testing. So in this experiment, we can get the goal of measuring concentration of droplets by this method.

3   MEASUREMENT SET-UP AND METHOD

3.1 Measurement set-up

An experimental system has been designed for measuring concentration of droplets based on ultrasonic method. The schematic diagram was shown in Figure 2: A humidifier generated the droplets which then passed through the plexiglass pipe, where two pairs of ultrasonic sensors were mounted on the side of pipe, to perform the pitch-catch experiments. Ultrasonic waves were generated by ultrasonic sensors working with the Signal Generator (arbitrary waveform generator), which could be adjusted to a tone-burst mode producing sine waveforms with 10-20 cycles at frequencies of 25 kHz and 40 kHz respectively. Ultrasonic signal captured by receiving sensors was converted into electrical signal and then sent to the computer by a data acquisition card after the necessary signal processing. The figure in Figure 3 has demonstrated an interface of the data acquisition and analysis program.

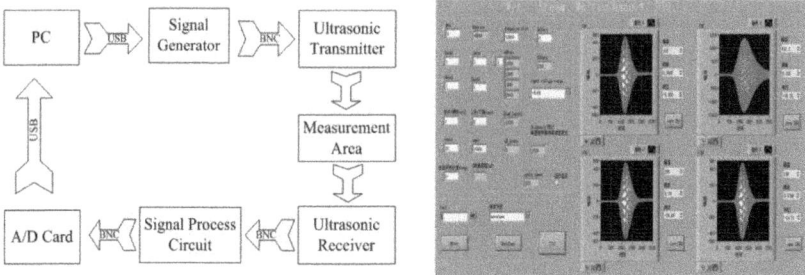

Figure 2. *Experimental schematic diagram*   **Figure 3**. *DAQ system in the experiment*

3.2 Measurement method

Firstly, ultrasonic sensors should be mounted on the plexiglass pipe and fixed in the opposite sides. Then the ultrasonic signal for the air (without any droplet) was acquired by the A/D card as the background which would be used the subsequent ultrasonic attenuation calculation. After that, adjust the humidifier to generating sufficient droplets steadily passing through the measurement zone. It should be noted that only when the vapor flow was steady, the ultrasonic signal could be recorded and saved. The ultrasonic attenuation coefficient could be calculated by comparing the amplitude variation with and without the droplets in the measurement zone (see Eq.(3)). Finally, by means of the prediction based on the ECAH model, the relationship between ultrasonic attenuation and concentration of droplets was obtained in this experiment, which truly established a direct connection between experimental signals and the unknown concentration. In addition, the results obtained by ultrasonic method with results obtained by light extinction method [9].

4   RESULTS AND DISCUSSION

4.1 The calculation of ultrasonic attenuation coefficient

The ultrasonic attenuation coefficient α_s in the gas-liquid two-phase flow was calculated as follows: L was the sound path indicating the distance between transmitters and receivers. A_0 and A_1 denoted the acoustic wave amplitude received by the sensors in the air (as the background) and in the circumstance of droplets passing through the measurement area (as the measurement signal) respectively. In Eq.(3), α_0 denoted the sound absorption coefficient in the air at normal temperature, which was looked up in reference [10].

$$\alpha_s = \frac{ln(A_0/A_1)}{L} + \alpha_0 \tag{3}$$

As shown in Figure 4, the amplitudes of signals decreased obviously when the droplets passed through the measurement zone, comparing with the signal amplitude in the air. To realize a fast testing, the data obtained from A/D combined with DAQ software designed by the author were processed using FFT technique and also analyzed by the software using LabVIEW, which ultimately yielded the concentration of droplets.

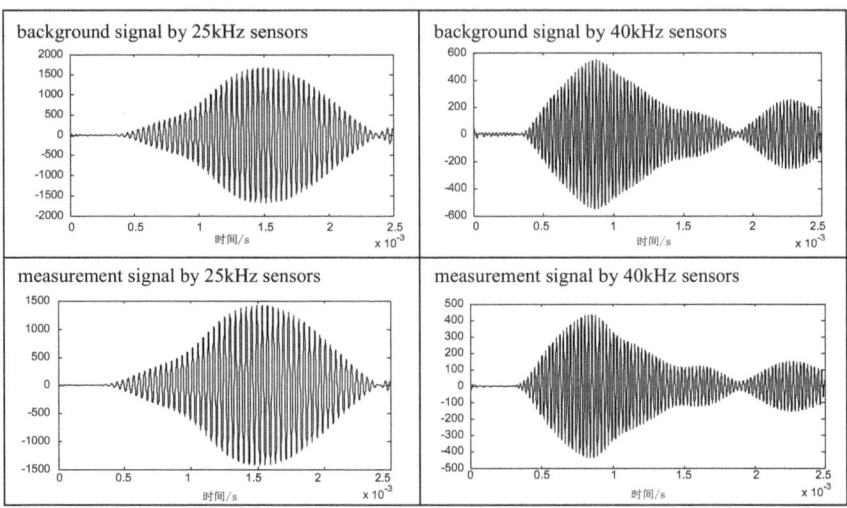

Figure 4. *Comparison of background signal and measurement signal in the experiment*

4.2 Ultrasonic measurement results and discussion

Table 2 lists the ultrasonic amplitudes, attenuations and the measured volume concentration. It can be found that the ultrasonic attenuation coefficient at 40 kHz can be larger than that at 25 kHz, which is consistent with the numerical computation by different frequency and the volume concentration is less than 0.01%.

Table 2 *The volume concentration of droplets measured by ultrasonic method*

	f=25kHz	f=40kHz
L(mm)	81.8	81.8
A_0(mv)	1673.2	551.7
A_1(mv)	1421.0	431.3
α_s(Np/m)	2.006	3.031
C_V(%)	0.00715	

Table 3 *The volume concentration of droplets measured by light extinction method*

L(mm)	40.5		
	0.00642	0.00889	0.00885
Cv(%)	0.00751	0.00824	0.00775
	0.00826	0.00909	0.00836
average Cv(%)	0.00815		
Deviation	12.27%		

As mentioned before, the light extinction measurement has been carried out to verify the accuracy of the measurement about concentration of droplets in the air by ultrasonic method, which gave a value of 0.00815% for volume concentration of droplets in the air. It could be seen that results measured by ultrasonic method is close with the results of the light extinction method, with a deviation of 12.27%.

5 CONCLUSIONS

In this paper, the ultrasonic attenuation coefficients in the gas-liquid two-phase flow were measured by low-frequency ultrasonic sensors at 25 kHz and 40 kHz . The concentration of droplets was obtained based on the ECAH model. The light extinction experiment was also carried out to verify the reliability of ultrasonic attenuation method. A deviation of 12.27% is found between the results of two methods.

In view of the great significance of on-line measurement for the wet steam concentration by ultrasonic attenuation method, increasing more frequencies could be a good way to measure wet steam in that more information of ultrasonic attenuation can be obtained, and reduced errors and more reliable results can be expected too.

ACKNOWLEDGMENTS

This work was supported by the National Science Foundation(51176128，51076106)、 Shanghai Science and Technology Commission(10540501000)、Shanghai Education Commission(12ZZ142) and the Innovation Fund Project For Graduate Student of Shanghai(JWCXSL1301).

References

1 Cai X S, Su M X, Shen J Q. Measurement technology and application of particle size characterization. Beijing: Industry of chemical publishing house, 2010. (in Chinese)

2 Su Mingxu, Cai Xiaoshu, Xue Minghua, Dong Lili, Xu Feng. Particle sizing in dense two-phase droplet systems by ultrasonic attenuation and velocity spectra[J]. Sci china Ser E-Tech Sci, 2009, 52(6): 1502-1510.

3 Epstein P.S, Carhart R.R.. The Absorption of Sound in Suspensions and Emulsions, I. Water Fog in Air. J. Acoust. Soc. Am., 1953, 25 (3), pp. 553-565.

4 Allegra J.R., Hawley S.A.. Attenuation of Sound in Suspensions and Emulsions: Theory and Experiments. J. Acoust. Soc. Am., 1972, 51, pp. 1545-1564.

5 Ning De-liang, Gao Lei, Liu Xin-quan,Research and comparison of methods of steam wetness measurement, Journal of Engineering For Thermal Energy And Power,2009,24(2).(in Chinese)

6 Cai Xiaoshu, et al. A   Novel Integrated Probe System For Measuring The Two Phase Wet Steam Flow In Steam Turbine. Journal Of Engineering Thermal Science, 2001, 10(2).

7 Su M X, Shen J Q, Xu F, Cai X S. A modified method in calculating attenuation spectra for particle system . Technical Acoustics . 2010 , 29(6), pp. 1-4. (in Chinese)

8 McClements D J. Ultrasonic Characterisation of Emulsions and Suspensions. Advances in Colloid and Interface Science, 1991, 37,pp .33-72.

9 Zheng Gang, Sun Hao, Huang Tinglei, Yu Xianhuang. On-line Particle Concentration Measurement using Two Wavelengths Light Extinction. Chinese Journal of Scientific Instrument, 2000, 21(5),pp.533-535.

10 Du Gonghuan, Zhu Zhemin, Gong Xiufen. Acoustics .Nanjing University Press, 2012. (in Chinese)

NUMERICAL SOLUTION OF DYNAMICS OF PM$_{10}$ SUBJECTED TO STANDING-WAVE ACOUSTIC FIELD

X.F. YANG, F.X. FAN, M.J. ZHANG

Institute of Particle and Two-phase Flow Measurement, School of Energy and Power Engineering, University of Shanghai for Science and Technology, Shanghai 200093, China

1 INTRODUCTION

PM$_{10}$ whose aerodynamic diameter is 10 microns or less is one of the primary pollutants in the atmosphere. These particles, especially PM$_{2.5}$ with an aerodynamic diameter of no more than 2.5 microns, do great harm to both the ambient environment and human health [1-3]. Coal-fired power plants and industrial processes are important sources of PM$_{10}$. In order to control the particulates from these air polluting sources, dust removal devices, such as electrostatic precipitators, bag filters and wet scrubbers, are installed. Although the devices have achieved efficiencies higher than 95% on basis of particle mass, their efficiency in capture PM$_{10}$ is still not high enough[4,5]. Therefore, new approaches are required to control the particle emission. Acoustic agglomeration in which a sound wave is used to promote agglomeration of the particles is recognized as one of the main preconditioning technique to separate the fine particles from the flue gas stream[6,7]. Through acoustic agglomeration the fine submicron particles are enlarged to big micron-sized ones, thus the particles can be removed by the existing dust removal devices.

Single particle dynamics is a basic phenomenon in the processes of acoustic particle interaction and agglomeration. Decades ago, Brandt, Freund and Hiedemann[8] proposed a theoretical formula (BFH equation) for the entrainment motion of a particle in an acoustic field. Since it is based on too many simplifying assumptions, the range of validity of BFH equation is unknown. Gucker and Doyle[9] undertook experiments on the vibration amplitudes of droplets with a diameter ranging from 1.6 to 3.9 μm, aiming at determining particle size using acoustic field. Temkin and Leung[10] obtained theoretically a general oscillation velocity of a sphere subjected to an acoustic wave, but the velocity equation is too complicated to apply directly. Czyz [11,12] derived analytical expressions of the drifting velocity and number concentration change rate of particles in the acoustic field. However, Czyz's work focused only on particle motion under the influence of so called drift forces. Dodemand et al.[13] examined the forces acting upon particles in an acoustic suspension application with the density ratio of the particle to fluid being comparably small. As the gas density is much lower than the density of PM$_{10}$, the conclusions of their research can not be generalized to acoustic particle agglomeration applications. Aboobaker et al.[14] derived the trajectory and concentration equations of fine particles in the acoustic and flow fields to study the fractionation and segregation of fine particles. With the development of measuring technique, Hoffmann et al.[15,16] performed visualization experiments on trajectories of polydispersed irregular quartz particles of less than 50 μm and monodispersed glass beads of 8.1 and 7.9 μm in the standing-wave acoustic field. It is obvious that the particles used in these experiments are too large to represent the $PM_{2.5}$ particles. Zhao et al.[17] experimentally examined the trajectories of coal-fired $PM_{2.5}$ in a standing acoustic field (frequency: 3000 Hz, intensity: 138 dB) and showed that the particle oscillates and drifts in the meantime under the effect of standing acoustic field. Wang et al.[18] observed trajectories of $PM_{2.5}$ under the effect of plane acoustic field (frequency: 1000 Hz, intensity: 110 dB and 120 dB) by experimental methods, but the property of the acoustic wave (travelling wave or standing wave) is not given.

The researches above focused on analytically solving the differential equations of particle motion and experimentally observing and recording the particle trajectories. Obviously, complex differential equations can only be solved in a few cases. In addition, due to the limitation of measurement and control techniques it is difficult to fully understand the behavior of PM_{10} in the acoustic field through experiments. As computer technology develops rapidly, numerical simulation has become a powerful tool to analyze micro-scale processes. Numerical simulation of forces, motion and collision of PM_{10} in the acoustic filed has been carried out by Yuan et al.[19-22]. However, with a deeper understanding of natures of acoustic fields and properties of PM_{10}, it is found that the models of forces they used, which include only the pressure gradient force, are too simplified, and the particle densities adopted (1.18 kg/m^3 and 1000 kg/m^3) are very different from actual PM_{10} from industrial processes. Fan et al.[23] numerically investigated the particle motion and agglomeration under the effect of drag force in an acoustic field. Although practical values of PM_{10} parameters like density of 2000 kg/m^3 and diameter of 3 μm were applied to study the particle trajectory, this study is not sufficient to understand the motion of PM_{10} covering a wide range in size. Zhang et al.[24-26] proposed a collision efficiency model and simulated the macroscopic effects of acoustic agglomeration using quadrature method of moments and improved sectional algorithm, respectively. Nevertheless, the microscopic motion of the particles is not involved. Wang et al.[27] built a dynamical model of an aggregate produced by agglomeration of two sphere particles on the basis of the theoretical models in literatures [18] and [22]. However, in their work the information on forces and velocities was not presented.

Since so far the reports on numerical solutions of PM_{10} dynamics in the acoustic field still have deflects in force analysis and parameter selection, in the following section a new dynamical model of a single PM_{10} with a diameter of 0.1~10 μm in an acoustic filed is presented. The numerical solution to the equations of particle motion is carried out by employing the self-organized computational program based on the variable time-step fourth-order Runge-Kutta method combined with the second order implicit Adams interpolation method. The forces acting upon the PM_{10} as well as the velocities and displacements of the PM_{10} are obtained.

2   MATHEMATICAL MODEL AND NUMERICAL ALGORITHMS

2.1   Assumptions of the Numerical Model

In the present research considered are PM$_{10}$ particles suspended and flowing in a gas at a constant speed. An intense acoustic field is applied perpendicular to the gas flow. We focus on single particle dynamics under the effects of the acoustic field, thus the particle motion in the direction of gas flow and the particle gravity are not taken into consideration. In order to simplify the simulation, the following model assumptions are made:
(1) The gas is ideal and the gas flow is laminar.
(2) The particles are rigid spheres and no interactions of particles occur.
(3) Since PM$_{10}$ particles are in the micron or submicron size range, it is reasonable to assume that the gas phase and the sound wave are not influenced by the particles.
(4) The acoustic wave is a plane, one-dimensional standing wave created by two oppositely propagating sinusoidal traveling-waves of equal frequency and intensity.

2.2   Equations of the Acoustic Wave

Suppose that the wave propagation is in *x*-direction. The gas velocity induced by the standing-wave acoustic field can be written as

$$u_{gx} = u_a \sin(kx)\sin(\omega t) \tag{1}$$

where u_{gx} is the gas oscillation velocity at position *x* and time *t*, u_a is the maximum gas velocity, *k* is the wave number $k = \omega/c$, ω is the circular frequency, $\omega = 2\pi f$, *f* is the frequency and *c* is the speed of sound.

The maximum gas velocity u_a depends on the intensity of the traveling-waves which produce the standing wave and can be expressed as

$$u_a = 2\sqrt{2I/(\rho_g c)} \tag{2}$$

where ρ_g is the gas density and *I* is the intensity of the traveling-waves, which has a dimension of W/m^2. In practice, the acoustic intensity level *L* (dB) is more frequently used to describe the acoustic intensity, which is defined by

$$L = 10\log_{10}(I/I_0) \tag{3}$$

where I_0 is the reference intensity, whose value is 10^{-12} W/m^2.

2.3   Dynamical Model of a Single Particle

When an acoustic wave is employed to the gas-particle suspensions, the drag force, Basset force, pressure gradient force and virtual mass force almost act immediately on the particle. Hence, we get the equation of particle motion in x direction

$$m_p \frac{du_{px}}{dt} = F_{dx} + F_{bx} + F_{px} + F_{vx} \tag{4}$$

where m_p is the particle mass, u_p is the particle velocity, F_d is the drag force, F_b is the Basset force, F_p is the pressure gradient force, F_v is the virtual mass force and subscript x means component in x direction.

The drag force can be easily obtained by Stokes equation [28]

$$F_{dx} = 3\pi\mu_g d_p \left(u_{gx} - u_{px} \right) / C_c \tag{5}$$

where μ_g is the dynamic viscosity of the gas, d_p is the particle diameter, C_c is the Cunningham slip correction coefficient used to account for non-continuum effects when calculating the drag force on small particles, which is given by

$$C_c = 1 + Kn[1.257 + 0.400\exp(-1.10/Kn)] \tag{6}$$

where Kn is the Knudsen number, and $Kn = 2l_m/d_p$, in which l_m is the mean free path of the gas molecule.

The Basset force can be expressed as [13]

$$F_{bx} = \frac{3}{2} d_p^2 \sqrt{\pi\rho_g \mu_g} \int_0^t \frac{(\dfrac{du_{gx}}{dt'} - \dfrac{du_{px}}{dt'})}{\sqrt{t-t'}} dt' \tag{7}$$

The pressure gradient force acting upon the particle is expressed by [29]

$$F_{px} = m_g \left(\frac{\partial u_{gx}}{\partial t} + u_{gx} \frac{\partial u_{gx}}{\partial x} \right) \tag{8}$$

where m_g is the mass of the gas with a volume equal to the particle.

The virtual mass force can be given by [30]

$$F_{vx} = \frac{1}{2} m_g \left(\frac{\partial u_{gx}}{\partial t} + u_{gx} \frac{\partial u_{gx}}{\partial x} - \frac{du_{px}}{dt} \right) \tag{9}$$

Rearranging Eqs. (5), (6), (8), (10) and (11) yields the following equation of particle velocity

$$(m_p + 0.5m_g) \frac{du_{px}}{dt} = 3\pi\mu_g d_p (u_{gx} - u_{px}) / C_c + \frac{3}{2} d_p^2 \sqrt{\pi\rho_g\mu_g} \int_0^t \frac{(\frac{du_{gx}}{d\tau} - \frac{du_{px}}{d\tau})}{\sqrt{t-\tau}} d\tau$$
$$+ 1.5m_g \left(\frac{\partial u_{gx}}{\partial t} + u_{gx} \frac{\partial u_{gx}}{\partial x} \right) \tag{10}$$

The equation of particle displacement X_p can then be obtained by

$$\frac{dX_p}{dt} = u_{px} \tag{11}$$

2.4 Numerical Algorithm

Considering the particle velocity given by Eq. (10), it is natural to try to get numerical approximate solutions using an accurate and relatively fast numerical algorithm such as variable time-step fourth-order Runge-Kutta method, since this method can be stable for suitable time step sizes with a reasonable accuracy. After the particle velocity at a time step is obtained, based on Eq. (11) the particle displacement can be readily obtained by the second-order implicit Adams interpolation method, which is of the form

$$X_p(t + \Delta t) = X_p(t) + \frac{\Delta t}{2} [u_{px}(t + \Delta t) + u_{px}(t)] \tag{12}$$

where Δt is the time step size.

To solve Eq. (10), the Basset force needs to be computed. By breaking the general temporal velocity variations of the gas and the particle into a series of step changes, for instance, at time $t_0 = 0$ there are changes Δu_{gx0} and Δu_{px0}, at time t_1 changes Δu_{gx1} and Δu_{px1}, at time t_2 changes Δu_{gx2} and Δu_{px2}, and so on, the Basset force can be written as follows:

$$F_{bx}(t_n) = \frac{3}{2} d_p^2 \sqrt{\pi\rho_g\mu_g} \int_0^{t_n} \frac{(\frac{du_{gx}}{dt'} - \frac{du_{px}}{dt'})}{\sqrt{t_n - t'}} dt' = \frac{3}{2} d_p^2 \sqrt{\pi\rho_g\mu_g} \sum_{i=0}^{n-1} \frac{\Delta u_{gxi} - \Delta u_{pxi}}{\sqrt{t_n - t_i}} \tag{13}$$

The particle trajectory is attained based on the linear displacement in each computational time step, and the forces can be computed from Eqs. (5), (7)~(9) once the solutions to particle velocity and displacement are obtained. The properties of the gas, acoustic field and particle used in the numerical simulation are shown in Tab. 1.

Table 1 *Properties of gas, acoustic field and particle*

Temperature /°C	Static pressure /Pa	Acoustic intensity /dB	Particle density /(kg/m³)	Original position of the particle
25	101325	150	2000	$3\lambda/8$

3   RESULTS AND DISCUSSION

3.1   Comparison between Numerical Simulation and Theoretical Prediction

In cases of a small particle with high density suspended in a low density fluid subjected to an acoustic field, the particle can be entrained by the acoustic wave in the wave propagation direction. The entrainment motion of a particle is given by the well known BFH equation as follows

$$H = \frac{1}{\sqrt{1+(\omega\tau)^2}}$$
(14)

where H is the entrainment coefficient which is defined as the ratio of displacement amplitude of the particle to that of the gas at the same original position, and τ is the relaxation time given by

$$\tau = \frac{\rho_p d_p^{\,2} C_c}{18\mu_g}$$
(15)

where ρ_p is the particle density.

In this section, a numerical simulation is conducted under the condition that the dimensionless relaxation number $\omega\tau$ ranges from 0.1 to 100 by varying the acoustic frequency and the acoustic entrainment coefficients of a particle with a diameter of 1.0 μm are obtained. The numerical results are compared with the theoretical values determined by the BFH equation as shown in Figure 1, which shows that the numerical solutions are in good agreement with the theoretical results. However, the mathematical model and the computational algorithm in this work obviously give more information on forces, velocity and displacement during the particle motion process than BFH equation.

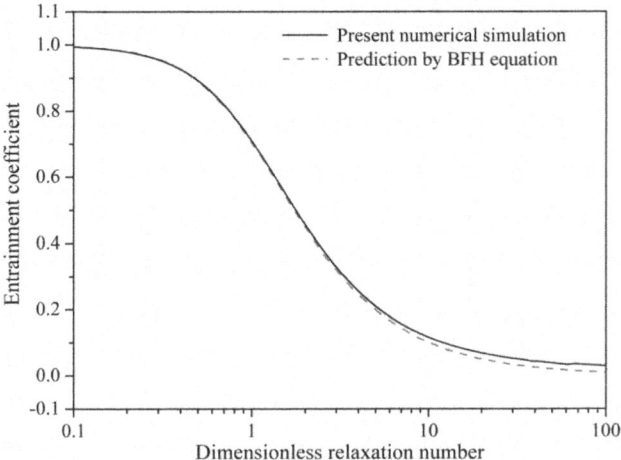

Figure 1   *Entrainment coefficient obtained from numerical simulation and comparison with theoretical prediction*

3.2   PM$_{10}$ Dynamics in the Acoustic Field

3.2.1   Forces Acting on PM$_{10}$.

Figure 2 gives the dependences of forces on time during the motion process of a particle with a diameter of 1.0 μm at the acoustic frequency of 1000 Hz. The simulation results show that in the initial stage with acoustic effect time of less than 0.033 ms, the drag force, Basset force and pressure gradient force increase rapidly from zero to their maximums, and the virtual mass force decreases dramatically to zero. It is noted that after the initial stage all of the forces change periodically with time in a sine curve pattern, and that the frequency of the forces is in accordance with the acoustic frequency. The duration time of the initial stage represents the time needed for the particle motion to become stable. It corresponds to 4.6 times the relaxation time of the particle (τ=7.1 μs calculated by Eq. (15)). For a practical acoustic agglomeration application, the particle will stay in the acoustic field for several seconds, thus the unstable initial stage can be neglected compared with the residence time. It is also noted that the magnitudes of the forces differ greatly. The drag force of the order of 10^{-12} N is the dominant acting force, whereas the Basset force, pressure gradient force and virtual mass force in the order of 10^{-14} N, 10^{-15} N and 10^{-16} N respectively can be neglected compared with the drag force in this computational case.

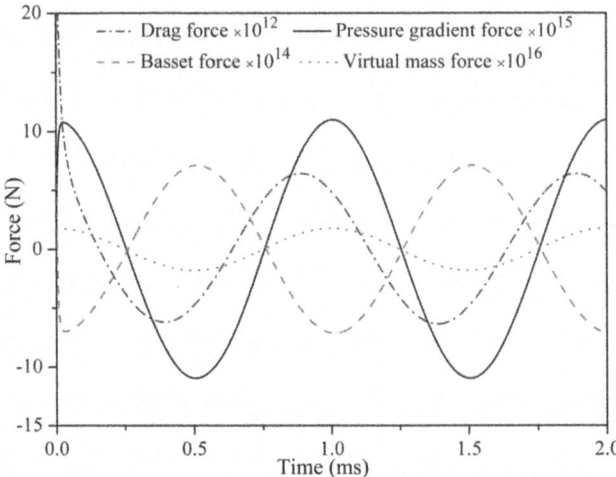

Figure 2   *Changes of forces with time during the motion process of a particle*

Since the forces acting on the moving particle change with time in a sine curve pattern, we then can focus on the force amplitudes to examine the influences of acoustic frequency and particle diameter on them. The changes of force amplitudes with frequency in the motion process of a particle of 1.0 μm in diameter are shown in Figure 3. It can be found that as the acoustic frequency increases, the force amplitudes increase. In the double logarithmic coordinates, the relationships between the force amplitudes and frequency approximate to linearity. The rates of increase in the drag force and pressure gradient force with an increase in the acoustic frequency are almost equal and smaller than that in the Basset force and virtual mass force.

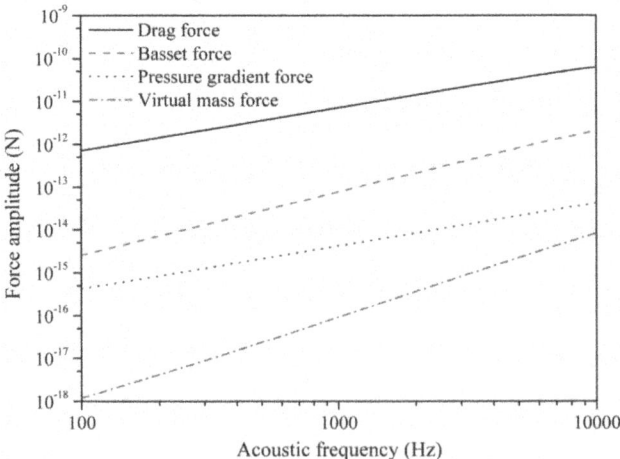

Figure 3   *Changes of force amplitudes with frequency during the motion process of a particle*

The forces acting on the particle are also affected by the particle size as shown in the equations of forces in Section 2. The changes of force amplitudes with particle diameter under an acoustic frequency of 1000 Hz are given in Figure 4. It is noted that the amplitudes of forces increase with an increase in particle diameter. It is also can be seen that as the particle diameter increases the difference in magnitudes of Basset force and drag force decreases, and that the amplitude of virtual mass force gradually approaches that of the pressure gradient force as a result of its high rate of increase.

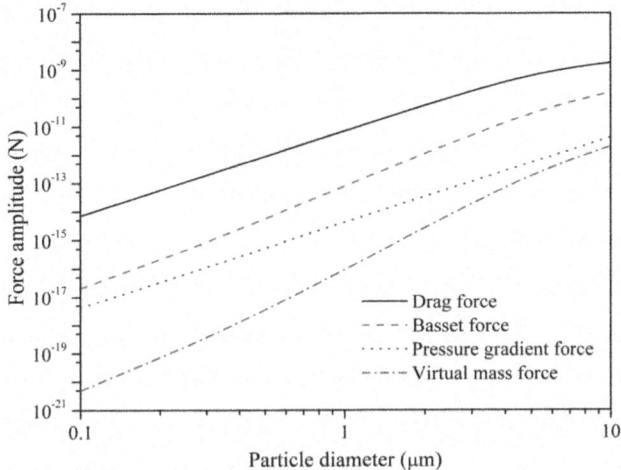

Figure 4   *Changes of force amplitudes with particle diameter during the motion process of a particle*

3.2.2. Velocity of PM$_{10}$ in the Acoustic Field.

The velocities of a particle of 1.0 μm in size under the effect of acoustic field at typical frequencies are given in Figure 5. We can see that the velocity changes periodically with time and that the effect of acoustic frequency on particle velocity appears mainly in the period of velocity. As the drag force plays a dominant role in determining particle motion (Figure 2), the particle velocity period is equal to the acoustic period, which decreases with an increase in acoustic frequency. For the fine particle, the velocity amplitude almost does not change with the frequency as shown in Figure 5. It has been found in Figure 3 that the amplitude of force increases with an increase in acoustic frequency. The increasing force amplitude leads to the increase of velocity, whereas the decreasing force period hampers the increase of velocity. Therefore, the velocity amplitude almost keeps constant in these cases due to the joint effects.

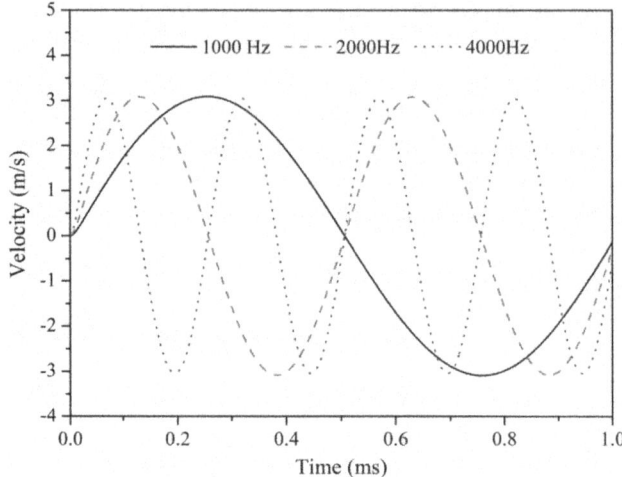

Figure 5   *Dependence of particle velocity on time for typical frequencies*

Figure 6 presents the dependences of particle velocity on time for typical particle diameters in cases with an acoustic frequency of 1000 Hz. It is shown that the particle diameter is an important factor influencing the particle velocity. Fine particles with a diameter of 1.0 μm or less can be fully entrained due to their low inertia, thus they have nearly the same velocity in the motion process. The velocities of micron sized particles differ significantly. The bigger the particle is, the smaller the velocity amplitude. It is also shown that the phase of particle velocity is different for the micron particles. The phase lag increases with the particle size. The reason is that a bigger particle responds more slowly to the periodical gas velocity field due to its higher inertia.

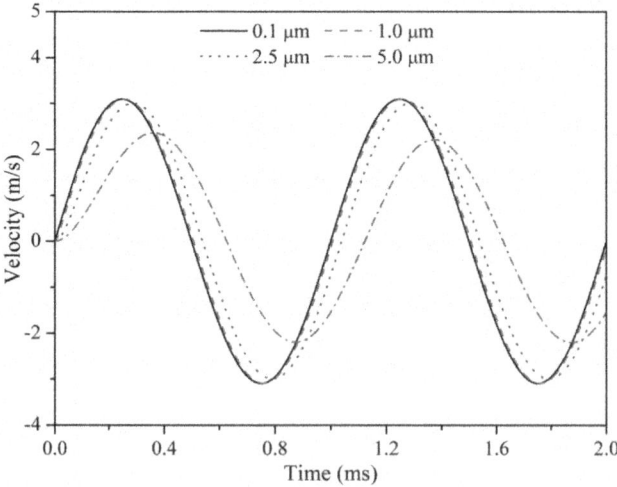

Figure 6   *Dependence of particle velocity on time for typical diameters*

The effects of the acoustic frequency on the velocity amplitude of PM$_{10}$ for different particle diameters are given in Figure 7. It is noted that as the acoustic frequency increases, the velocity amplitude of the particle with 0.1 µm in diameter keeps constant. In cases with an acoustic frequency lower than 2000 Hz, the velocity amplitude of the particle with 1.0 µm in diameter remains unchanged; however, when the acoustic frequency is higher than 2000 Hz, it begins to decrease with the frequency. The higher the frequency is, the higher the rate of decrease in particle velocity amplitude. It is also noted that the velocity amplitude of the particle with a diameter of 2.5 µm begins to decrease at an acoustic frequency of 400 Hz and that the particle with a diameter of 5 µm shows a monotonously decreasing behavior in the velocity amplitude with an increase in acoustic frequency.

To further explain the dependence of the particle velocity on the acoustic frequency and particle diameter, a parameter commonly used to measure particle inertia in a fluid is introduced. It is defined as the ratio of the relaxation time of the particle to the characteristic time of the flow field[31]. For the gas-particle suspension system under the effect of acoustic field, the wave period (1/f) can be adopted as the characteristic time of the gas phase. Therefore, the particle inertial parameter Ω is given by

$$\Omega = \tau f = \frac{\rho_p d_p^2 f C_c}{18 \mu_g} \tag{16}$$

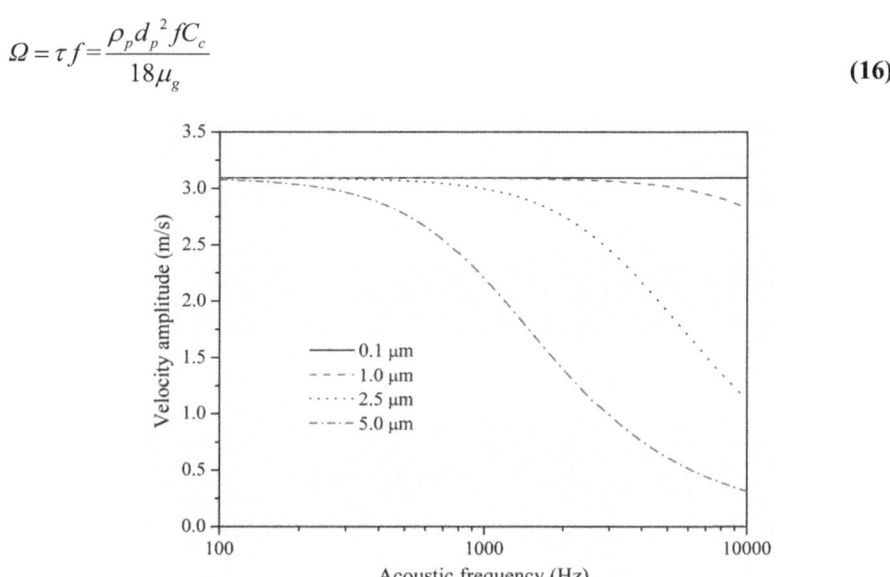

Figure 7 *Dependence of particle velocity amplitude on frequency for different particle diameters*

In case with an acoustic frequency of 100 Hz, the particle no more than 5.0 µm in diameter belongs to inertialess range ($\Omega \ll 1$), thus it can be fully entrained by the oscillating gas. Therefore, the velocity amplitudes of these particles are equal as shown in Figure 7. In cases with acoustic frequency of 10000 Hz or lower, for the particle with a diameter of 0.1 µm, the inertial parameter is of the order of 10^{-3} or less, hence the particle can be considered as inertialess particle, resulting in a constant velocity amplitude in Figure 7. For the micron particle of 5.0 μm in size, $0.015 < \Omega < 1.57$, the relationship of $\Omega \ll 1$ is not satisfied, therefore the particle is referred to as inertial particle and the particle velocity is significantly influenced by the frequency.

3.2.3 Displacement of PM₁₀ in the Acoustic Field.

The displacements of a particle of 1.0 μm in size with time for different frequencies are given in Figure 8, where the positive displacement indicates that the particle moves on one side of its original position. This motion behavior can be explained by Figure 5. For instance, let us consider the case with an acoustic frequency of 1000 Hz. As shown in Figure 5 for the first half of an acoustic cycle the particle velocity increases from zero with time at first, after reaching its maximum it decreases and returns to zero. In this process the particle moves continuously towards to the positive direction of the *x*-axis and reaches the maximal displacement as shown in Figure 8. For the second half cycle, the velocity changes to the opposite direction, increases from zero to its maximum, and then decrease to zero (Figure 5), as a result the particle moves towards and finally returns to the original position.

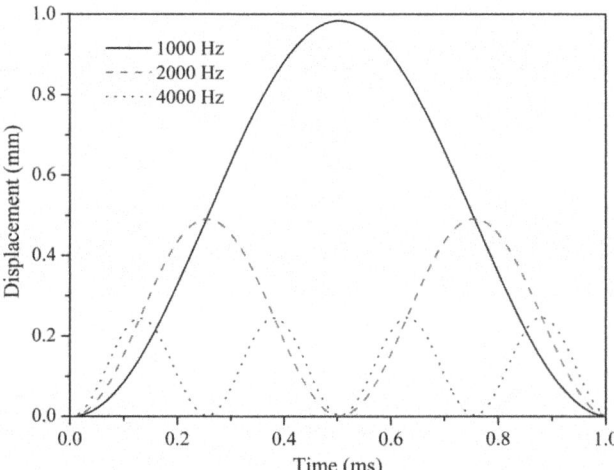

Figure 8   *Dependence of particle displacement on time for typical frequencies*

The displacements of particles with different diameters change with time for an acoustic frequency of 1000 Hz are shown in Figure 9. It is noted that there is sufficient similarity in trajectories of particles of 0.1 and 1.0 μm in sizes. The particles go to their displacement peaks then reverse to the original position, and the cycle repeats. It is also noted that particles of 2.5 and 5.0 μm in sizes return to their neutral positions which are not the original position after an initial acoustic cycle, and that the greater the particle size is, the bigger the distance between the new neutral position to the original position. The particle motion behavior is determined by the particle inertia. Once the acoustic field is applied, the inertialess particle no greater than 1.0 μm in size almost immediately acquires a velocity equal to the carrier gas velocity, thus the particle has enough time to reach the point where the motion started in an acoustic cycle. In case with a large micron particle whose inertia is significant, an appreciable period of time is needed for the particle to achieve oscillating equilibrium in the gas-particle system, then the particle does not have enough time to follow the oscillating gas in the first acoustic cycle, as a result, the neutral position differs from the original position.

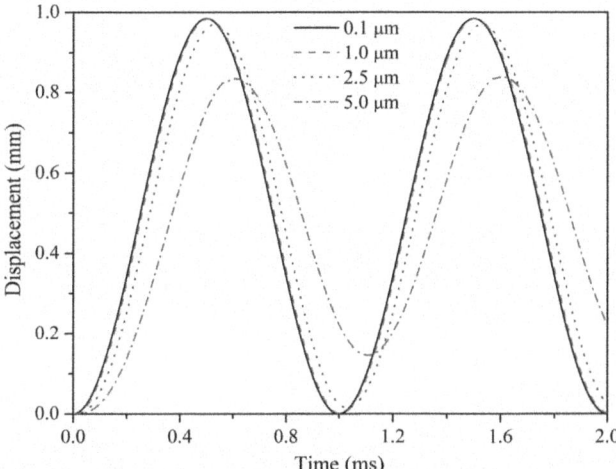

Figure 9 *Dependence of particle displacement on time for typical diameters*

Figure 10 shows the dependence of the displacement amplitude of PM$_{10}$ on the acoustic frequency for different particle diameters. Here, we can see clearly that the displacement amplitude decreases as the acoustic frequency increases. The reason lies not only in that the displacement amplitude of the oscillating gas decreases with an increase in acoustic frequency, but also in that the acoustic particle entrainment coefficient decreases as a result of the increase of particle inertial parameter with the acoustic frequency. Also, it can be seen that in cases with an acoustic frequency lower than 300 Hz, the particle displacement amplitude curves practically coincide. As the acoustic frequency increases further, the displacement amplitude curve for a particle of 5.0 μm becomes to separate from the curves for the particles no more than 2.5 μm at the frequency of 300 Hz, and the curve for a particle of 2.5 μm separates from the curves for the particles no more than 1.0 μm at the frequency higher than 1000 Hz. However, the curves for particles of 1.0 μm and 0.1 μm appear to overlap till the frequency of 10000 Hz. These results can be well explained by the inertia of the particle using the inertial parameter given by Eq. (16).

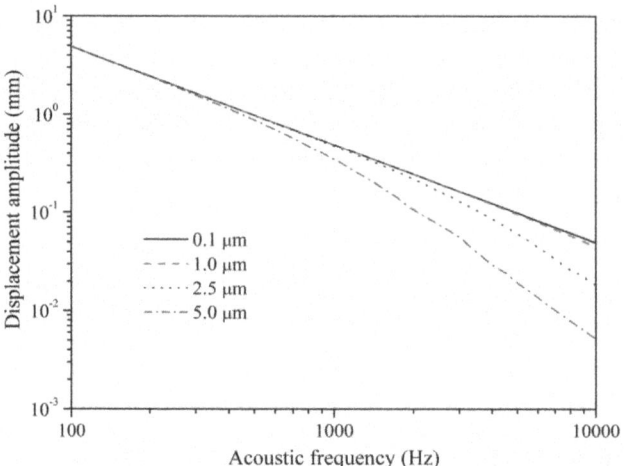

Figure 10 *Dependence of particle displacement amplitude on frequency for different particle diameters*

4   CONCLUSION

The dynamics of a single suspended PM_{10} under the effect of acoustic standing wave has been elucidated by mathematical models. The particle motion equations are numerically solved using the variable time-step fourth-order Runge-Kutta method combined with the second-order implicit Adams interpolation method. This approach results in rather accurate numerical solutions with entrainment coefficients very close to the analytical solutions predicted by the BFH equation. The particle dynamic parameters, like the force, velocity and displacement in the wave propagation direction of the applied acoustic field during particle motion process, are simulated. Among the drag force, Basset force, pressure gradient force and virtual mass force, the drag force plays a dominant role in determining particle trajectory. The motion of the particles changes with the acoustic frequency. When the frequency is very low, the particles of different sizes behave in a very similar way. As the frequency increases, the particle motion begins to differ from each other. In addition, significant differences are observed in the velocity and displacement of submicron and micron sized particles. An inertial parameter is introduced to characterize the inertia of a particle in the gas-particle suspensions. According to this parameter, PM_{10} in the gas flow subjected to an acoustic field can be in the inertialess or inertial range, depending mainly on the acoustic frequency and particle size.

References

1   Y. W. Niu, L. Y. He, M. Hu, J. Zhang, Y. L. Zhao, *Sci. China, Ser. B. Chem*, 2006, **49**, 466.

2   R. Meij, B. te Winkel, *Fuel Process. Technol.*, 2004, **85**, 641.

3   S. S. Park, A. S. Wexler, *J. Aerosol Sci,* 2007, **38**, 509.
4   W. Peukert, C. Wadenpohl, *Powder Technol.,* 2001, **118**, 136.
5   Q. Yao, S. Q. Li, H. W. Xu, J. K. Zhou, Q. Song, *Energy,* 2009, **34**, 1296.
6   T. L. Hoffmann, *Ultrason.,* 2000, **38**, 353.
7   J. Z. Liu, G. X. Zhang, J. H. Zhou, J. Wang, W. D. Zhao, K. F. Cen, *Powder Technol.,* 2009, **193**, 20.
8   O. Brandt, H. Freund, E. Hiedemann, Z. *Phycik,* 1937, **104**, 511.
9   F. T. Gucker, G. J. Doyle, *J. Phys. Chem.,* 1956, **60**, 989.
10   S. Temkin, C. M. Leung, *J. Sound Vibr.,* 1976, **49**, 75.
11   H. Czyz, *Arch. Acoust,* 1987, **12**, 199.
12   H. Czyz, *Acustica,* 1990, **70**, 23.
13   E. Dodemand, R. Prud'homme, P. Kuentzmann, *Int. J. Multiphase Flow,* 1995, **21**, 27.
14   N. Aboobaker, D. Blackmore, J. *Meegoda, Appl. Mathl Model,* 2005, **29**, 515.
15   T. L. Hoffmann, G. H. Koopmann, *J. Acoust. Soc. Am,* 1996, **99**, 2130.
16   I. González, T. L. Hoffmann, J. A. Gallego, *J. Aerosol Sci,* 2000, **31**,   1461.
17   B. Zhao, G. Yao, X. L. Shen, *Proc. CSEE,* 2007, **27**, 13.
18   C. B. Wang, Q. Li, H. Chen, *Proc. CSEE,* 2007, **27**, 18.
19   Z. L. Yuan, W. L. Li , X. Wei, F. X. Fan, X. L. Shen, *Proc. CSEE,* 2005, **25**, 121.
20   Z. L. Yuan, W. L. Li, X. Wei, F. X. Fan, X. L. Shen, *J. Southeast Univ,* 2005, **35**, 140.
21   Z. L. Yuan, F. X. Fan, G. Yao, B. Zhao, X. L. Shen, *J. Combust Sci. Technol,* 2005, **11**, 298.
22   F. X. Fan, Z. L. Yuan, *Proc. CSEE,* 2006, **26**, 12.
23   F. X. Fan, Z. L. Yuan, B. Zhao, G. Yao, *J. Combust Sci. Technol,* 2008, **14**, 253.
24   G. X. Zhang, J. Z. Liu, J. H. Zhou, K. F. Cen, *J. Chem. Ind. Eng,* 2009, **60**, 42.
25   G. X. Zhang, J. Z. Liu, J. Wang, J. H. Zhou, K. F. Cen, *J. CIESC,* 2011, **62**, 922.
26   G. X. Zhang, J. Z. Liu, J. Wang, J. H. Zhou, K. F. Cen, *J. Combust Sci. Technol,* 2012, **18**, 44.
27   Z. Wang, X. H. Zhong, Y. Yan, R. Ge, *Chin. J. Environ. Eng* 2011, **5**, 2839.
28   Y. Nakajima, T. Sato, *Powder Technol,* 2003, **135-136**, 266.
29   B. Wang, A. B. Yu, *AIChE. J.,* 2010, **56**, 1703.
30   M. Simcik, M. C. Ruzicka, J. Drahos, *Chem. Eng. Sci.,* 2008, **63**, 4580.
31   V. M. Alipchenkov, L. I. Zaichik, *Fluid Dynam.,* 2001, **36**, 93.

DISCHARGE ANALYSIS OF AN INDUSTRIAL BATCH ROTATING DRUM

Y.S. Cheong[1], A. Zhao[1], H. Ahmadian[2], W. Bi[1], R. Shen[1]

[1] Procter and Gamble, Beijing Innovation Centre, No. 35 Yu'an Road, Tianzhu Konggang Development Zone B, Shunyi District, Beijing 101312, China.
[2] Procter and Gamble, Newcastle Innovation Centre, Whitley Road, Newcastle upon Tyne, NE12 9BZ, United Kingdom.

1 INTRODUCTION

In industrial applications, tumbling mixers are used extensively to perform mixing of particulate materials either in continuous or batch mode[1]. One type of tumbling mixer exists in the form of a cylindrical vessel rotating along the length of the vessel, refer thereafter as rotating drum. Rotating drums are normally operated in the cascading regime to allow mixing of particles. Published data suggested that mixing in the radial direction of the mixer is considerably faster than along the axial direction[2]. Therefore, rotating drums are normally incorporated with baffles to enhance axial mixing and thus, improving process efficiency.

In batch operations, particles need to be discharged following mixing and processed further downstream such as storage or packing. In contrast to mixing, the study of material discharge from a rotating drum has been scarce. The current work was aimed to provide insights into the mechanism and factors governing discharge using a combination of experiments and simulations. Discrete particle simulations were performed to facilitate visualization of particle flow field during discharge thus providing an understanding of the phenomena observed in the discharge experiments. This paper concludes with a proposal to improve the efficiency where batches are mixed and discharged consecutively by truncating the discharge cycle appropriately.

2 METHOD AND RESULTS

A pilot scale horizontal rotating drum with 5 baffles from Munson Machinery Inc. was used to perform the discharge experiments and the drum was filled with 500kg of detergent granules at the beginning of each trial. During discharge, the end plate covering the exit of the drum was tilted inwards and the drum was rotated at a set speed to encourage material flow out of the drum. A buggy placed on load cell was used to collect the materials and the weight of the buggy was recorded every 10s until the whole batch was completely discharged in order to generate discharge curves. Drum rotation speed was the only variable tested during the experiments and varied according to Table 2 covering approximately an order of magnitude difference in Froude Number.

In order to provide insights into the particle flow behaviour, discrete particle simulations were performed using the open-source code LIGGGHTS. The well-established Hertz-Mindlin contact laws[3] were used and the key simulation parameters employed are tabulated in Table 1. The geometry of the drum was generated based on the dimensions of the pilot scale mixer used in the aforementioned experiments.

Table 1 *Hertz-Mindlin parameters used in the discrete particle simulations.*

Poisson ratio (-)	0.25 (particle), 0.3 (wall)	Particle diameter (mm)	20(95%),40(5%)
Shear modulus (Pa)	1×10^8 (particle), 7×10^9 (wall)	Coefficient of restitution (-)	0.4
True density (kg/m³)	970 (particle), 7800 (wall)	Coefficient of static friction (-)	0.5

The discharge mechanism may be identified by inspecting the simulated particle flow depicted in Figure 1. It can be seen from the figure that particles were lifted by the baffles to the top of the mixer and avalanched out of the mixer due to the effect of gravity. On this basis, one may expect that the discharge time may be minimized by increasing the amount of materials being lifted either by increasing the rotation speed or the number of baffles. The effect of rotation speed will be considered next.

The total time required to completely discharge approximately 500kg at different drum rotation speeds were measured experimentally and tabulated in Table 2. It is evident that the discharge time increased with increasing rotation speed, which was in contrast to the expectation above. Three of the discharge profiles are shown in Figure 2 where the symbols are measured data and the solid lines are fits of a rate equation given by 1-exp(-Rt^n). R is a rate constant, t is the time elapsed and n is an empirical constant. At the rotation speeds investigated, the values of n were all approximately 1.4, the rate constants at 9, 12 and 19rpm were 0.008, 0.006, 0.004 s^{-1}, respectively. The reduction in rate constant with increasing rotation speed is consistent with the experimental data shown in Table 2. Such phenomenon can be explained by the simulated flow field in Figure 1(b) where high rotation speed caused more particles to accumulate in between the baffles at the top of the mixer. This may be a consequence of an increase in centrifugal forces acting on the particles with increasing rotation speed.

Inspection of Figure 2 suggests that approximately 90% of the batch weight was discharged during the first 75% of the total discharge time except at 19rpm when the centrifugal force was too high. When considering mixing then discharge for multiple batches in a consecutive manner, one can truncate the discharge cycle appropriately and feed in the next batch to start the mixing process in order to reduce the overall process time. Such approach will result in an increase in the fill level of next batch due to the material retained from the previous batch and hence, care should be taken such that adequate mixing can still be achieved without extending the mixing cycle with the new fill level.

Table 2 *Experimental test conditions and the measured times taken to completely discharge the batches.*

Rotation speed (RPM)	9	12	14	19	27
Froude Number (-)	0.075	0.133	0.181	0.333	0.673
Discharge time (s)	66	73	75	76	137

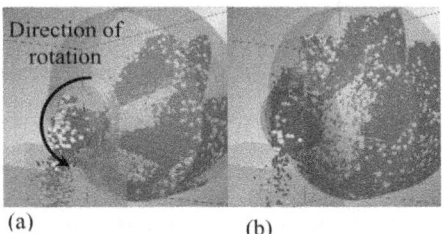

(a)   (b)

Figure 1 *Simulated discharge flows at (a) 9rpm and (b) 27rpm. Dark and light were used to represent 20mm and 40mm particles, respectively.*

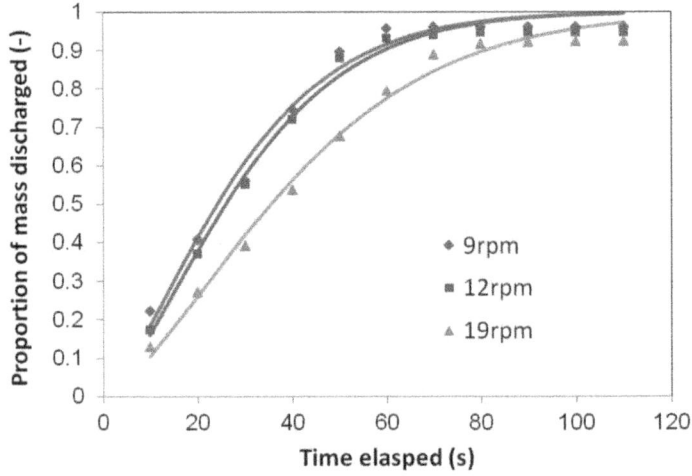

Figure 2 *Discharge profiles at different rotation speeds.*

3   CONCLUSION

In conclusion, the use of a combination of experiments and discrete particle simulations has provided powerful insight into the discharge process of a horizontal rotating drum. The discharge of particles was driven by the avalanche mechanism as particles were lifted by the baffles incorporated in the mixer. However, it was necessary to choose a rotation speed such that excessive centrifugal forces were avoided to minimize the discharge cycle time. Finally, the pros and cons in truncating the discharge cycle were discussed in the context of consecutive mixing-discharge of multiple batches.

References

1   L.T. Fan, Y.M. Lai, F.S. Lai, Recent development in solids mixing. *Powder Technol.*   **61** (1990) 255 – 287.
2   J.M. Ottino, D.V. Khakhar, Mixing and segregation of granular materials, Annu. Rev. *Fluid Mech.* 32 (2000) 55–91.
3   C. Thornton, K.K. Yin, M.J. Adams, Numerical simulation of the impact fracture and fragmentation of agglomerates, *J. Phys.* D 29 (1996) 424-435.

A COMPREHENSIVE TECHNOLOGY OF PARTICLE CHARACTERIZATION THAT AUTOMATICALLY MEASURE PARTICLE SIZE, SHAPE AND CHEMICAL IDENTITY IN ONE SINGLE PLATFORM

Brian Li

Malvern Instruments Ltd, Enigma Business Park, Grovewood Road,
Malvern,Worcestershire,UK
Brian.li@malvern.com.cn

1 INTRODUCTION

The Morphologi G3-ID (Figure 1) combines the capabilities of a Morphologi G3 particle analyzer withthe Kaiser RamanRxn 1 spectrometer resulting in a single platform measuring particle size, shape and chemical identity. The Raman spectrometer uses a 785 nm laser allowing the spectral range 150 cm^{-1} to 1850 cm^{-1} to be acquired with a 4 cm^{-1} spectral resolution. The spectrometer accessory is coupled using fiber optics to the Morphologi G3 allowing Raman spectra to be acquired. In addition to reporting the standard morphological parameters, the Raman capability on the Morphologi G3-ID enables the chemical identification of particulates in a dispersed sample. In a situation where particles in a multi-component sample all present a similar morphology, chemical information might be the only way to distinguish between the particles of interest for evaluation. The extra chemical information permits creation of chemical classifications from which component- specific particle size distributions are determined. This can provide extra compositional information about samples that can be invaluable in research and development or investigative situations.Here we will provide a case to show application in pharmaceutical industry. In order to show that a generic drug is bioequivalent to an innovator drug it must display comparable bioavailability when studied undersimilar experimental conditions[1]. Bioavailability is the rate and extent to which the active ingredient isabsorbed from a drug product and became available at the site of drug action and bioequivalence refers to equivalent release of the same drug substance from two or more drug products or formulations[2]. The premise underlying this 1984 law is that bioequivalent products are therapeutically equivalent and, therefore, interchangeable. Malvern Instruments received samples of innovator and generic tablets of a dual active pharmaceutical ingredient (API) product. The interest was investigation of particle sizedistribution (PSD) of each of the APIs upon tablet disintegration as this would be expected to have a significant effect on the subsequent bioavailability of the drug.

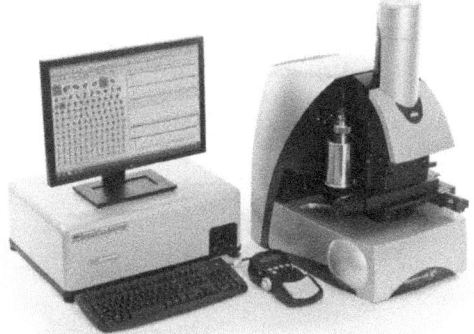

Figure 1: *Morphologi G3-ID*

2 METHODS

The tablets provided contained the same dosage for the two APIs. Both of which were practically insoluble in water. This was therefore used as the dissolution/disintegration media, with the presumption being that the API particles would remain mostly undissolved. One tablet of each was dissolved in 100 ml of water. 2 ml was sub-sampled and diluted with a further 20 ml of water. 2 ml of this suspension was pipetted onto an aluminium coated microscope slide and allowed to dry overnight. The particle size and shape data were collected and analyzed with automated image analysis, with settings determined and stored as a standard operating procedure. A spectral reference library was created for the sample by taking point spectra of the "pure" components, Figure 2. For this analysis, the size range of interest was between 1 and 10 µm and Raman spectra were acquired from only particles in this size range.

3 RESULTS

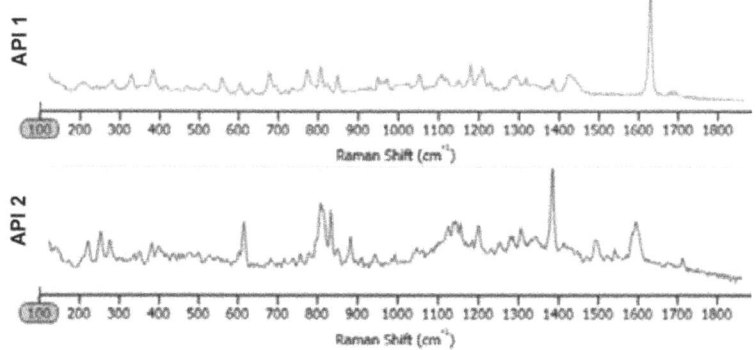

Figure 2: *Reference API spectra*

Based on an analysis of morphological parameters from the particle image results alone, the two APIs could not be differentiated. The particles were too similar in shape. The inclusion of Raman spectroscopic information for chemical identification readily differentiated between the APIs, as demonstrated in Figure 3. This scattergram plots the correlation scores for the two API components against one another. API 1 and API 2 are clearly separated via Raman data. Also shown in the figure are example particle images.

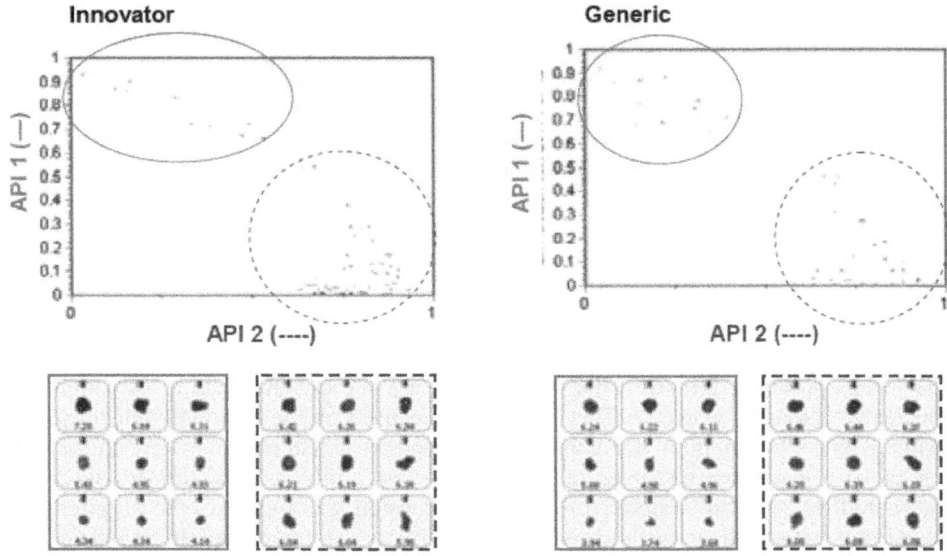

Figure 3: *Scattergram of API score values and example images from each class*

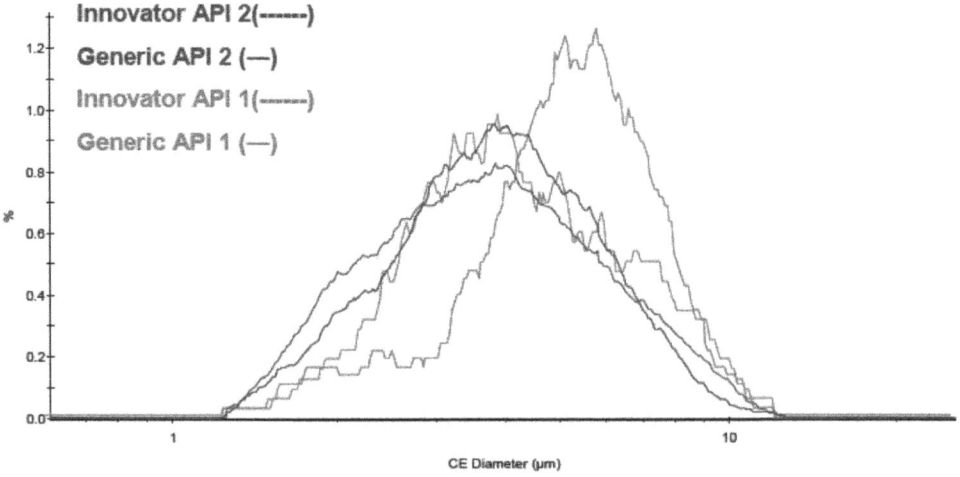

Figure 4: *Overlay of API PSDs for both tablets*

Once the two API populations can be differentiated, it then becomes possible to determine the individual PSDs from blends. Figure 4 shows the overlay of the circular equivalent diameter (CED) distribution by number of each API from the chemically defined populations. There appears to be fewer small particles of API 1 in the generic tablet than in the innovator tablet. The API 2 PSDs appear to overlay well for the two different tablets. Figure 5 and Figure 6 show classification charts comparing the two API classes for the two tablet types in percentage count and percentage volume, respectively. The tablets contain equal amounts of each API in their formulations, but the innovator appeared to contain a higher proportion of API 2 compared to API 1 than the generic tablet, for the samples analyzed.

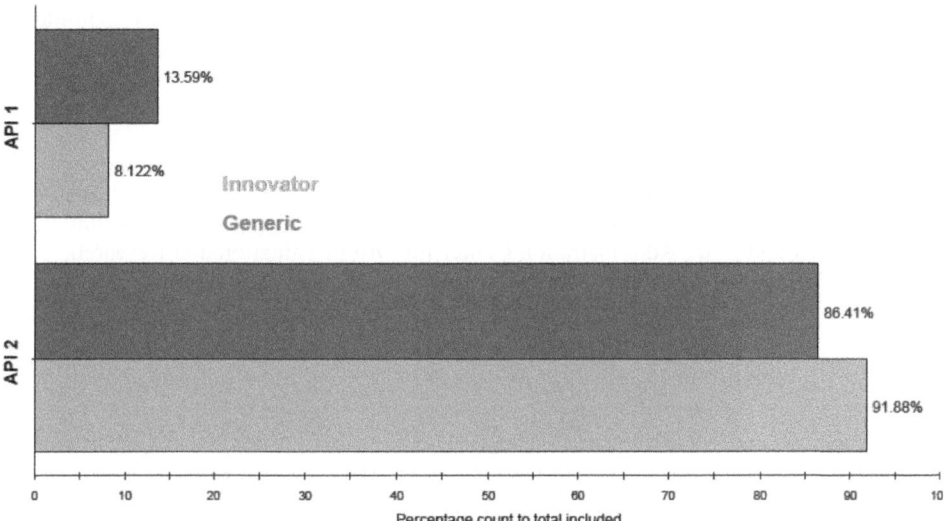

Figure 5: *Classification chart showing API classes by percentage count*

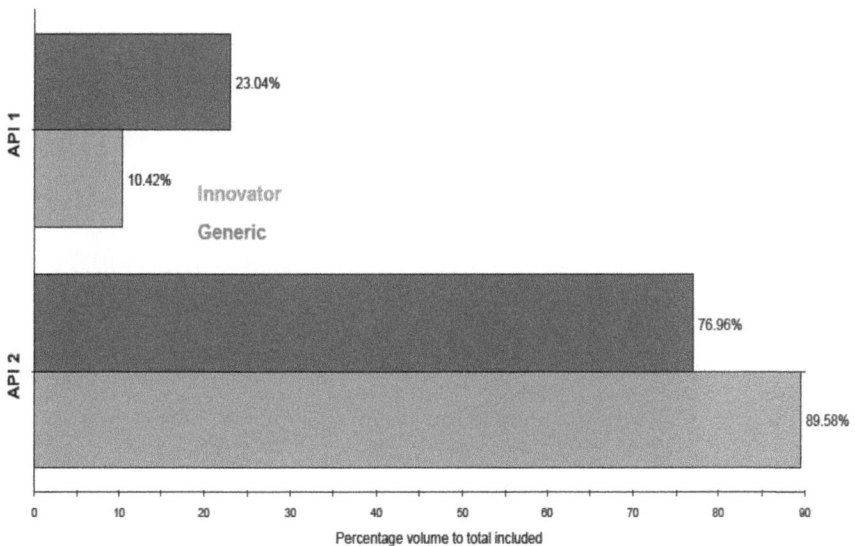

Figure 6: *Classification chart showing API classes by percentage volume*

4 CONCLUSION

The combination of automated particle imaging and Raman spectroscopy in one instrument allows Morphologically Directed Raman Microscopy to be performed. This allows the individual components present within a blend or mixture to be independently characterized and compared.

Such a tool can be used to gain better product understanding across many areas of pharmaceutical industry from regulatory to troubleshooting. It is not, however, limited to pharmaceuticals alone and is also applicable to other samples which have Raman spectra.

References

1   Federal Food, Drug and Cosmetic Act, section 505(j)(8).
2   "Orange Book Annual Preface. Statistical Criteria for Bioequivalence". Approved Drug Products with Theraputic Equivalence Evaluations 29th Edition, U. S. Food and Drug Administration Center for Drug Evaluation and Research, 18-06-2009.

AIR CURRENT SEGREGATION IN INSUTRIAL SILOS – A DESIGN CHALLENGE FOR FILTERS IN THE AIR EXTRACTION SYSTEM

R.J.Farnish[1*], S.Zigan[2], J.J.Rodriguez[1]

[1] The Wolfson Centre for Bulk Solids Handling Technology, University of Greenwich, Central Avenue, Chatham, Kent ME4 4TB, UK
[2] School of Engineering, University of Greenwich, Central Avenue, Chatham, Kent ME4 4TB, UK
* R.J.Farnish@gre.ac.uk

1   INTRODUCTION

The loading of storage silos with fine powders such as carbon dust requires a profound understanding of the segregation behaviour of the material in the silo. Segregation occurs because of difference in particle size i.e. fines (particles smaller than 40 microns) are behaving hydrodynamically different than larger particles because the drag force on the smaller particles is larger. This causes the separation of the fines from the coarser particles. The separation process is called air current segregation and occurs during the silo filling process. When the silo is filled the particles settling into the silo accelerate the air through transference of drag effects. The air starts circulating and the moving air creates local areas in the silo with different velocity and pressure characteristics. Fines entering a zone with higher air velocities are accelerated and carried away from the main bulk and often are sucked out of the silo by the extraction system[1]. This results in a significant pressure drop across the filter. The pressure drop across granular filters was studied experimentally by[2]. The pressure drop is strongly influenced by the filter material and the properties of the filter[3]. The fines start to build up in the filters until the filters are saturated with fines and the filter needs to be cleaned. The cleaning of the filters is a challenging process because with time the fines are accumulating in the filter membrane and filter efficiency reduces dramatically. Experiments were conducted using different filter types. The filter was loaded and cleaned in a continuous process and the filter efficiency was analysed.

2   METHOD AND RESULTS

2.1 Test rig configuration

A test apparatus has been assembled at The Wolfson Centre for Bulk Solids Handling Technology to evaluate pressure drop profiles associated with particle embedment in different types of filter media configuration (Figure 1). Key parameters controlled within the test rig were particulate feed rate, gas mass flow rate, filter media specification and cleaning technique. The test was designed to accommodate filter candles of up to 2m length and maximum diameter of 100mm. Gas flows to the test rig were directed via a choked flow nozzle bank operating at 7 bar (abs), local to the test rig a supplementary pressure regulator was installed (typically operating to 1 barg). Powder feeding into the gas flow was achieved through the use of a pressure balanced screw feeder which discharged into a 150mm blowing seal rotary valve. The controllable outputs of the screw feeder and gas flow enabled the test rig to operate with a range face velocities or solids loading ratio (SLR). Filter cleaning was achieved through the use of reverse jet pulsing using a pressurised gas reservoir configured to allow the use of a range of pressures. The bulk particulate used for the research was carbon black. The clean side of the filter system was connected to a flow interceptor (within which a removable adhesive impingement surface could be placed to examine carry over) en route to a high flow HEPA filter.

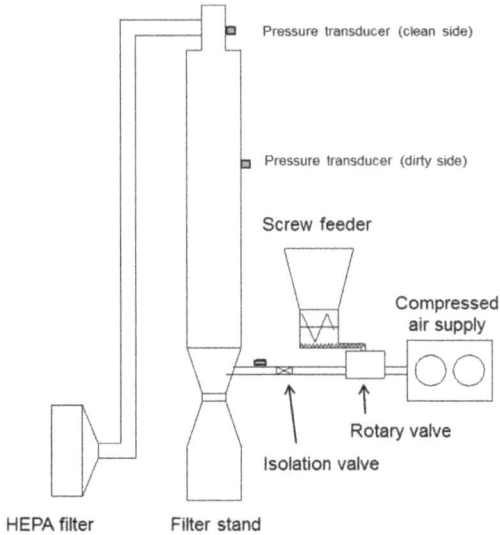

Figure 1 *Schematic diagram of the test rig*

The key purpose of the test rig was to evaluate the rate of increase in pressure drop in response to changes in (primarily) face velocity for different types of filter configuration (i.e. in-depth versus face filtration). More importantly for industry, the residual pressure drop following cleaning was examined (i.e. permanently embedded material). From the industrial perspective this latter aspect of work is complementary to primary investigation since filtration efficiency must be considered as a combination of particle removal, attainment of a low working pressure drop upon return service – but also ideally resilience to long term pressure degradation over time. Hence, the ability to capture particles is of secondary importance if the subsequent cleaning cycle cannot release material from within the volume of the filter media effectively enough to maintain an acceptable level of operation.

2.2 Method of operation

The design of the test apparatus was such that different modes of operation could be accommodated to support the initial investigative programme considering particle embedment and removal. The primary configuration utilised a pleated filter constructed from sintered media (7AL3) (Figure 2) and the test particulate was carbon black (Figure 3). The filter media has a flat sheet depth of ~48µm offers a 5µm particle cut off and was orientated on the series of filters to support either in-depth or face filtering requirements (the difference being defined as to whether the finer pore structure is innermost and the coarser pores outermost or vice versa respectively). The carbon black is not well defined in terms of particle size due to its propensity to readily form agglomerates (Figure 4) which is trait of such fine particles, however for the purposes of this investigation the median particle size is reported as having a d10, d50 & d90 of 2µm, 5µm & 180µm respectively. In considering how this size range might interact with a filter having a 5µm cut-off, it is worth noting that examination of a dust capturing adhesive pad place on the clean side of the filter showed no significant carry-over of fines (such as would be expected), however the presence of a very small number of impacted agglomerates measuring up to ~200µm in size were detected – along with some discrete ~120nm particles. Inspection of the HEPA filter downstream from the filter showed no apparent deposits or darkening of the filter media to suggest that substantial fine carry-over had been occurring – even at the end of an extensive programme of trials. From these observations it is reasonable to assume that the particles contacting with the filter media during tests are likely to have been assemblies of particles greater than 5µm in size (i.e. the cut-off size for the filter media).

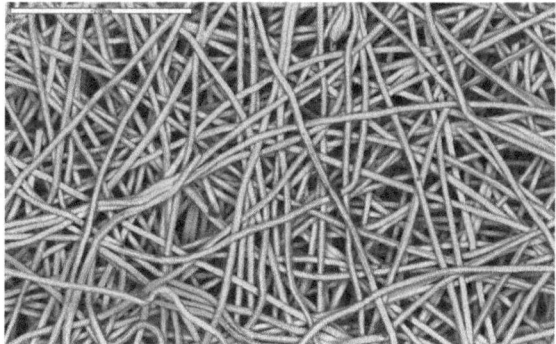

Figure 2 *SEM image of filter media (200µm scale bar, top left)*

Figure 3 *SEM image of the carbon black test media*

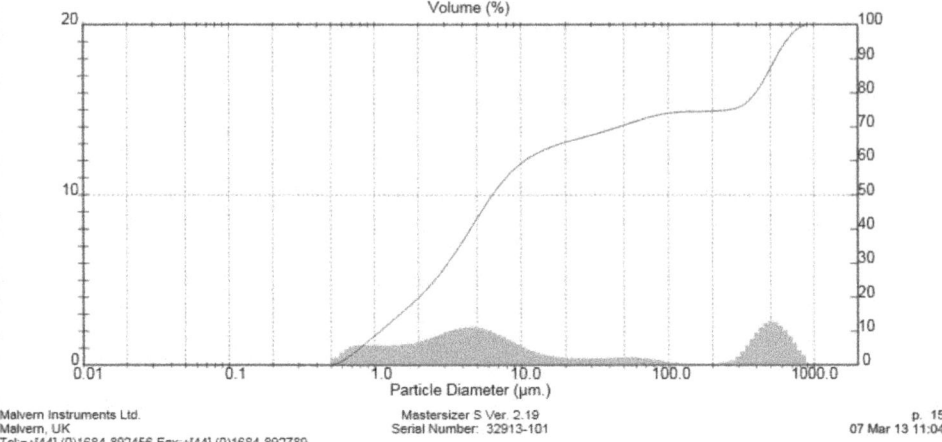

Figure 4 *Particle size distribution for the carbon black grade used in the trials – note secondary 'peak' around 400μm which reasonable be taken to represent agglomeration. Data: Malvern laser diffraction method, dry dispersion.*

Measurement of the air only pressure drop for the test filter were undertake to provide a base line condition for the 'clean' state followed by a 'loading' trial designed to operate until a target pressure drop of 600 mbar was arrived at.(Figure 5). Over the duration of this trial it can be seen that a degree of instability in the pressure trace was apparent towards the end of the trial – which was considered to be either a function of flow instability in the powder feeder or the dislodgement of accumulated carbon from with the internal surfaces of the test apparatus. Subsequent modifications to the powder feeder to address its flow instability served to minimize this effect in subsequent trials. The target powder feed rate was 3g/m (which was calibrated off line).

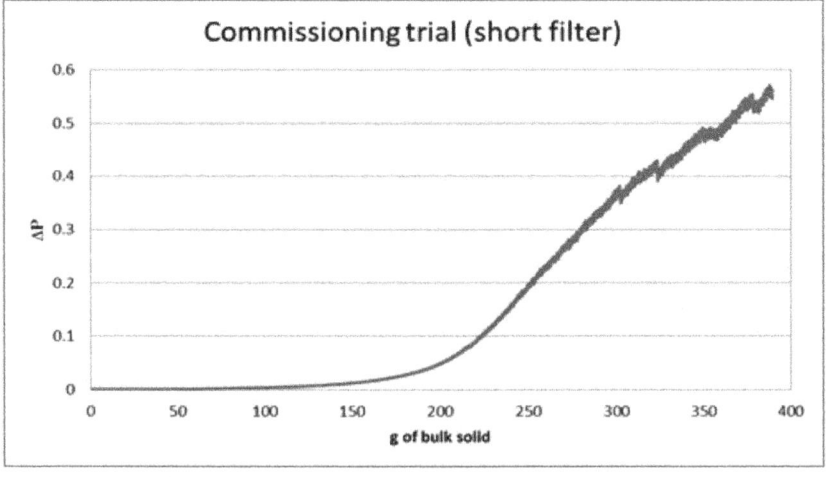

Figure 5 *Results from the initial loading trial*

Examination of the data obtained shows a very progressive increase in pressure drop from ~ 150g on the horizontal axis. It is considered that this early stage of the test represented a 'loading' phase whereby the carbon black adhered to various surfaces within the test apparatus and that, realistically; material transfer to the filter did not substantially increase until after this phase. With this in mind it would appear that it took approximately 300g of material to progress the filter to approximately 580 mbar pressure drop.

Subsequent to the loading trail, the filter was subjected to a series of tests to replicate an industrial installation. These tests maintained the previous powder feed rate of 3g/m and took the filter to a revised target maximum pressure drop value of 10mbar (a value derived from the particular requirements of the industrial co-sponsor of this research work).

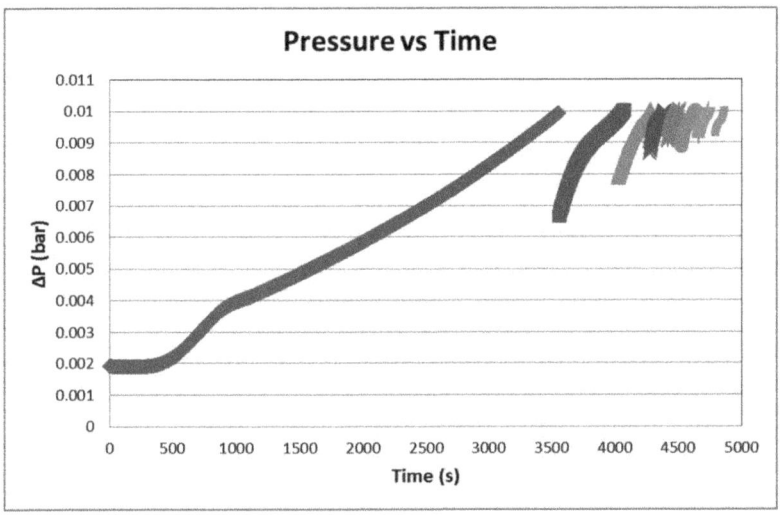

Figure 6 *Results from repeated loading and cleaning cycles for the filter*

Figure 6 presents the data obtained for this test series. The initial loading gradient exhibits the typically shallow approach to the peak pressure drop seen in all trials for in-depth and face filtering. Following the attainment of 10 mbar pressure drop, the filter was subjected to a reverse jet pulse (10 litres of air pulsed at a regulated 4.5 barg). Referring once again to Figure 6, it can be seen that the progression to the attainment of the trigger pressure has a pronounced non-linear characteristic compared to the initial loading gradient. Progressive cycling of the filter indicated that the 'forward' lean of the pressure trend characterized the loading of commissioned filters. This cycle of tests were undertaken until the filter was considered to have stabilized to an irretrievably 'dirty' condition.

Having taken the filter to a fully loaded condition (as defined by the parameters imposed for these tests), the filter was subjected to 10 reverse jet pulses and the cycle of tests re-initiated. Figure 7 shows the resulting set of data for this second series of tests.

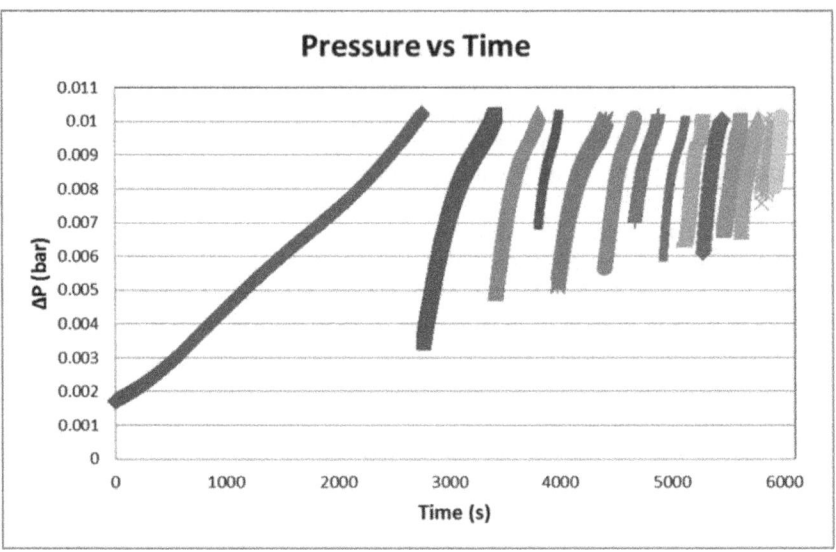

Figure 7   *Results from repeated loading and cleaning cycles for the filter*

The rate of initial loading for the filter was comparable to that of the filter in its unused condition (shown in Figure 6).   However, subsequent cycles of cleaning for the filter can be seen to have resulted in a lower 'cleaned' condition for all tests.   Also noteworthy in Figure 7 is the phenomenon of 'filter relaxation' evident after tests 4 and 6 in the series. This effect is evidenced by the return of the filter to a lower pressure drop if left in an undisturbed state.   In the case of the effect after test 4, this represented a 16 hour cessation of tests, whilst for test 6 this representation a test cessation of 1 hour.   Although this represents a poorly understood and seldom reported effect, it is clearly detectable in the instrumented test rig used for this study.

Repetition of the above test program for a face filter (i.e. fine pores outward and coarse pores inward), showed a steeper initial loading gradient, but still delivered the forward sloping trends for subsequent cycles.   In common with the in-depth filter, the second series of tests showed the development of lower pressure drop values for the 'cleaned' conditions – as well replicating the 'relaxation' effect.

3   CONCLUSIONS

The results presented in this paper represent a small element of a much larger investigation, but it is felt that the salient points of the work are well covered by the reported outputs.

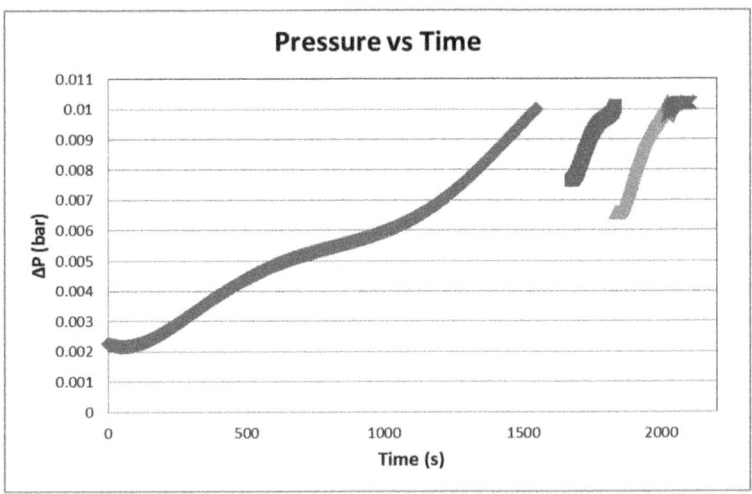

Figure 8 *Results from the initial series of repeated loading and cleaning cycles*

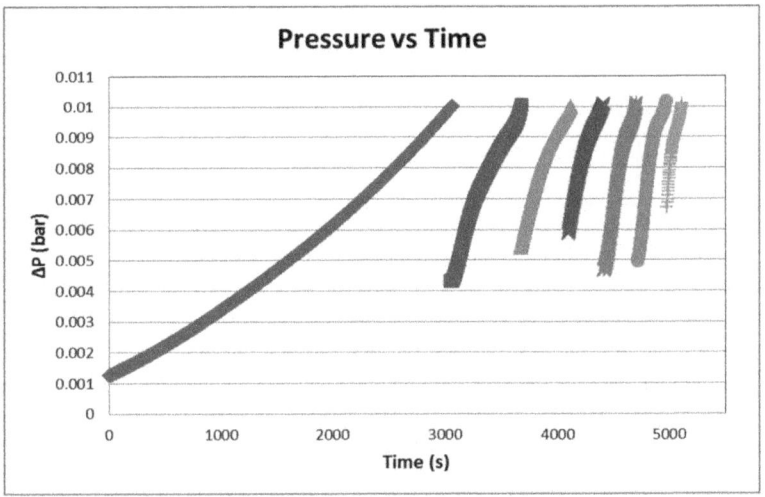

Figure 9 *Results from the second series of repeated loading and cleaning cycles*

The majority of work relating to filter operation has its basis in wet processes (vacuum, pressure or centrifugal) where the efficient and predictable operation of equipment for particle capture is key to the profitability and reliability of value adding processes. In such instances the deep penetration of particles is undesirable and filter design reflects the need for particle release from the surfaces of the selected media. Such work as has been undertaken usually considers particle capture (and subsequent filter efficiency) in terms of a) cake filtration, b) blocking filtration or, c) depth filtration[2]. Figure 10.

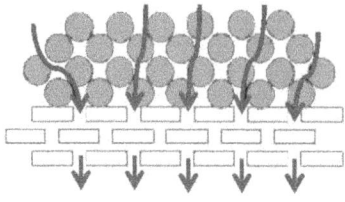

a) Cake filtration

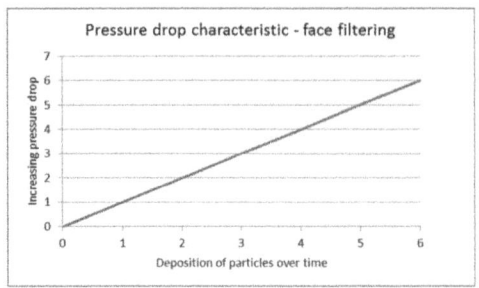

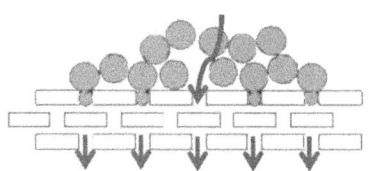

b) Blocking filtration

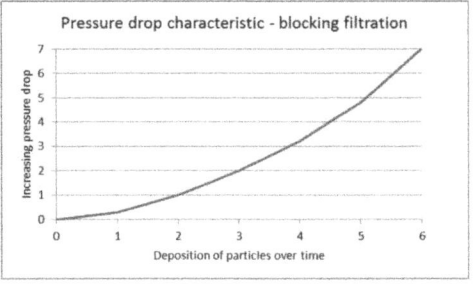

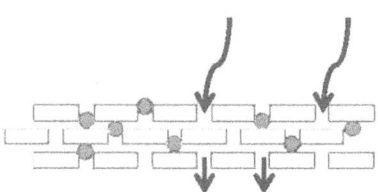

c) Depth filtration

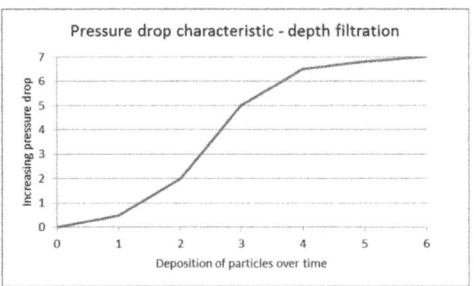

Figure 10   Illustrations to support the three main concepts for particle capture

In the case of cake filtration the assumption is made that the deposition of particles onto the surface of the media is homogenous in terms of particle size (or distribution) and porosity for the deposition area. If these conditions exist then, for a constant rate of particle deposition, the rate of increase in pressure drop is likely to be linear. For blocking filtration, the response of the system pressure drop is dictated by the effect of direct blocking of pores by particles (which is analogous to blinding of small aperture sieves when working with equally fine or soft particles). Depth filtration is generally taken as defining the adhesion of ultra-fine particle (substantially smaller than the filter fibre) to the strands forming the filter media through the depth of the membrane or the embedment of particles within open internal flow channels. This approach to describing mechanisms of particle is useful as a basis for considering approaches to modeling the behavior of particles captured in dry systems.

The key difference between wet and dry systems lies in the dominance of the viscosity of the carrier medium relative to the particle properties. In this respect drag forces tend to dominate the mechanism of particle accumulation or capture – which has the implication that for comparable particle behavior significantly higher face velocities would be required in dry systems if an appropriate Reynolds number is to be generated.

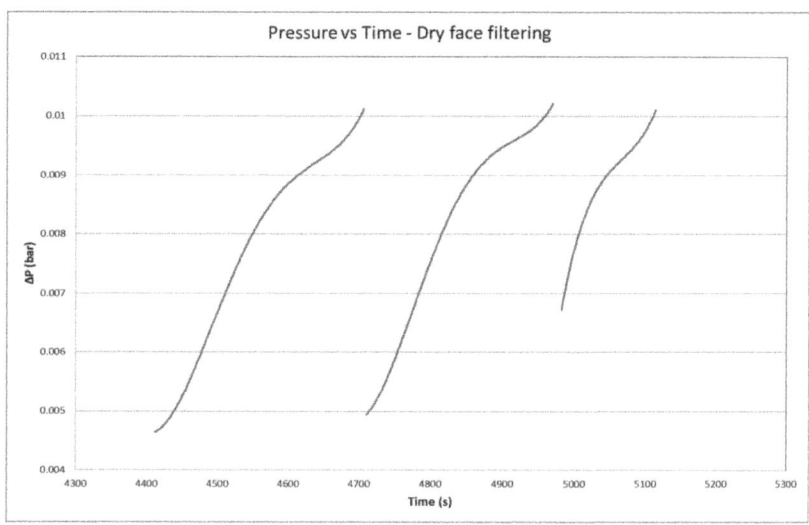

Figure 11 Pressure profiles for face filtering for a solids-gas system (first loading cycle)

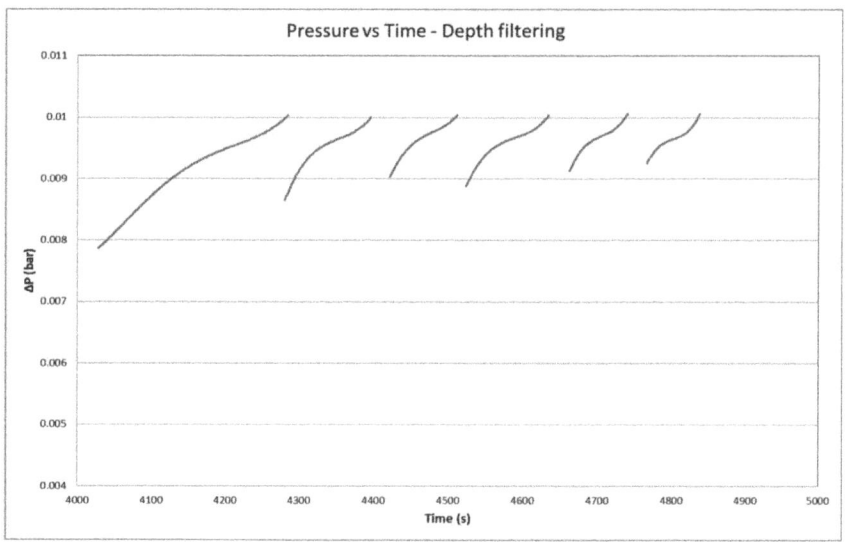

Figure 12 Pressure profiles for depth filtering for a solids-gas system (second loading cycle)

Considering the pressure profiles for both face and depth filtering for dry solids (Figures 11 &12), it is apparent that the characteristic of the profile does not correlate well with the accepted definitions for reducing filter performance. Most noticeable is that the trends for both types of filter exhibit a compound form that is highly repeatable. It could be argued that the mechanism that is developing within the filters that have been tested is a compound of depth filtering initially which evolves into blocking filtration (Figure 13). This serves to illustrate that a difference approach would be required to predict filter performance for the types evaluated to date. It is quite likely that a 'one model fits all' approach may not be attainable due to the number of variables involved in filter operation – most significant of which would be the construction of the filter media and the particle characteristics.

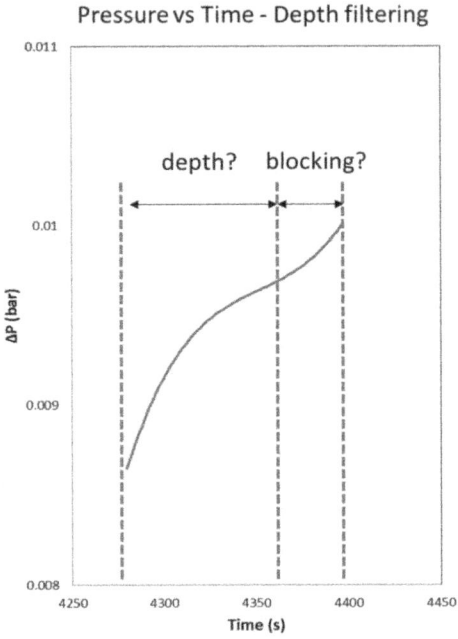

Figure 13   Possible compound effects for reduced filter performance over a single cycle for a depth filtering trial – using existing terms of reference.

Research into dry systems has also been undertaken, but has been primarily focused on fine particulate capture (e.g. control of diesel emissions[2] or aerosol capture[3]), and only consider irreversible particle entrapment – not industrial cyclic operations over an extended period.

Considering the data obtained from the preliminary work reported upon in this document. The following model is proposed:

Taking a clean filter as an example to consider particle embedment, it can be appreciated that initially the ΔP values will be both low and consistent over the whole available surface area of the candle. It is likely that this would result in higher initial contact loadings over the sections of the candle first exposed to the incoming flow of suspended particles. As these areas begin to load with material (resulting in a locally higher ΔP), the gas flow and its entrained particles will progressively distribute over the available area. During this redistribution phase it is possible that cumulative higher ΔP would result in higher face velocities at the start to the particle loading in the lower ΔP regions – resulting in deeper penetration effects. These conditions may be transitional only during bringing a new or 'cleaned' filter back into service on station. Once 'steady state' conditions are established, the development and nature of ΔP progression could change to a more macroscopically influenced mechanism linked to flow path migration through the media. As particles become drawn into the media by the gas flow, it is reasonable to assume that the initial gas flow channels to become established will be those with either the most direct path or having the most consistent large (relatively) cross sectional area – both conditions representing a low ΔP passage. The larger paths may be more susceptible to the adhesion of very fine material within the confines of the channel are progressively choke the route. Smaller paths will be inherently vulnerable to direct blocking. It is plausible that the population of low ΔP channels through the media will become exhausted relatively early in the life of the filter (although how early cannot be readily estimated). Subsequent back flushing is may serve to reactivate a proportion of the most favourably size/routed channels, but the less direct channels that would have been 'forced' by the gas in the intervening period will be progressively more difficult to flush free of particles. Considering mechanism within the pleats it possible that the proximity of the media faces at the 'apex' of the pleat could result in the interaction of incoming gas flow to generate a region of increased ΔP compared to the bottom of the pleat where the media surfaces are significantly wider apart. Thus during early operation of the filter, higher face velocities will be attained towards the base of the pleat – resulting in firmer embedment and progressively weaker embedment towards the apex of the pleat.

If this model is taken as adequately (and accurately) identifying some of the key factors at work in industrial scale operations, it can appreciated that the application of face filtration (cake deposition) models or even a hybrid model incorporating in-depth (blocking) models cannot provide a useful approach to predicting filter performance.

Clearly, the development of an understanding of the physical behavior of filtration systems has not obtained the prominence in research directed work that has been achieved by wet processes over the decades. In having highlighted and identified several important characteristic of dry filter systems, it is anticipated that the development of the research facilities at The Wolfson Centre for Bulk Solids Handling Technology will serve to redress this situation and ultimately provide operational data and models applicable to equipment suppliers and end users alike.

References

1   S. Zigan, R.B. Thorpe, U. Tuzun, G.G. Enstad, F. Battistin, *Theoretical and experimental testing of a scaling rule for air current Segregation of alumina powder in cylindrical silos, 2008,* **183,** 133.

2   A.J. Torregrosa, J.R. Serrano, F.J. Arnau, P. Piqueras, *A fluid dynamic model for unsteady compressible flow in wall-flow diesel particulate filters,***36,**671.

3   M. Masoudi, *Hydrodynamics of diesel particulate filters*, SAE technical paper 2002-01-1016, 2002.

4   W. Gosele, *Ullmann's Encyclopedia of Industrial Chemistry,* 2000, Filtration, **1-65.**

A COMPARATIVE STUDY ON THE INFLUENCE OF PARTICLE SIZE ON THE TURBULENCE CHARACTERISTICS WITHIN GAS-SOLIDS PNEUMATIC FLOWS USING AN ELECTROSTATIC SENSOR AND CFD-DEM COUPLED SIMULATION

Jianyong Zhang[1], Wei Chen[2], Ruixue Cheng[1], Kenneth Williams[2], Mark Jones[2] and Bin Zhou[3]

[1]University of Teeside, United Kingdom
[2]Centre for Bulk Solids and Particulate Technologies, The University of Newcastle, Australia
[3]Southeast University, China

1. INTRODUCTION

In pneumatic conveying, the flow rate of particles often needs to be monitored or controlled. Accurate measurements of flow velocity, concentration and flow rate are critical[1]. Electrostatic, electrical capacitance and microwave are three common methods used in such applications[2,3]. Among them, electrostatic meters measure the charges carried by particles by detecting the induced charge or voltage on the electrode[4,5]. Figure 1 demonstrated the principle of this technique.

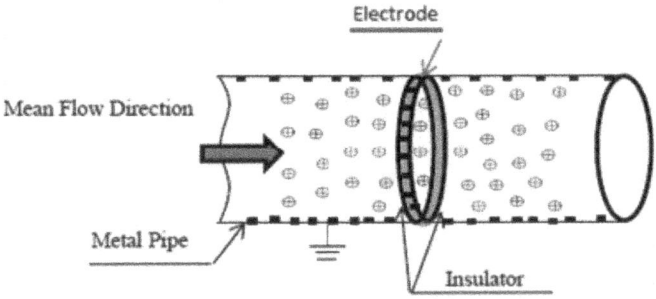

Figure 1 *Principle of the Electrostatic sensor.*

For convenience of explanation, assume that particles carry positive charges (in reality, particle can be negatively charged too), the charge induction occurs at the inner surface of the earthed metal pipe wall and the insulated floating metal electrode. The conditioning circuit is used to detect the charge induction on the electrode only. Indirectly, the output of the conditioning circuit can be used to indicate solids flow.

In industrial environment, electrostatic sensors are susceptible to low frequency noises. Hence, dynamic measurement is usually used, which measures the fluctuation of the charge or voltage induced on the electrode. It is this feature that makes "dynamic" electrostatic meters sensitive to the level of flow turbulence.

In single phase flow, the flow turbulence level is measured by Reynolds number which is proportional to its mean velocity if the viscosity of fluid is given. In dilute gas-solids two-phase flows, to the authors' knowledge, there is no commonly accepted method to express the level of turbulence. In this paper, radial velocity to axial velocity ratio is used as an indicator. One of reasons for using this ratio is that the dynamic electrostatic meters can be used to detect the change of turbulence. This can be demonstrated using Figure 2. The fluctuation level of signal and its frequency band increase with its mean value, where the mean value of the signal is corresponding to the mean flow rate. So the fluctuation (root mean square value is often used) of signal has been used to indicate the mean flow rate. This is the fundamental assumption that dynamic electrostatic meters are based on.

However, due to the high conveying velocity in dilute gas-solids flows, particle degradation occurs from recursive particle-particle and particle-wall collisions. This results in: 1) Particle size reduction along with the conveying process. Therefore, the total number and total surface area of particles continuously increase for a given amount of material mass. 2) As total surface area increase and particle size become smaller, the surface tension overwhelms the gravitational force, the solid phase follows gas-phase more smoothly, and the turbulence level of particles reduces.

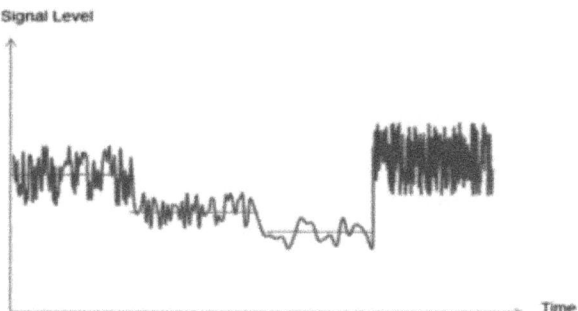

Figure 2. *Typical electrostatic sensor output.*

The above two phenomena have opposite effects to an electrostatic dynamic meter: on one hand, larger surface area leads to higher total charge carried by particles; however the turbulence level decrease results in smaller fluctuation on the other hand. The eventual signal trend (rise or fall) depends on the shift between these two effects. This paper presents simulation and experimental results which confirm each other regarding particle degradation and resultant turbulence level change.

2. EXPERIMENTAL SCHEME

The experiments were conducted using the Teesside University pneumatic conveying rig as show in Figure 3. The rig is comprised of one vertical and two horizontal sections. Four 40mm diameter electrostatic meters are installed along the rig. The draft fan sucks air in, which provides transport for solids fed from the screw feeder. Under the screw feeder, a weighing platform provides measurement of weight loss, from which the mass flow rate of solids is derived on line. The air-solids mixture passes through the horizontal and vertical sections. The signals picked up by four meters are acquired by the computer. The mean velocity of particles at each of the above four locations is measured using the cross correlation method, the signal induced on the electrode is also used for flow rate measurement. The solids and air are separated in the cyclone, and the air flow rate is also measured so that the system can provide on line air flow rate, particle flow rate and air to solids ratio simultaneously.

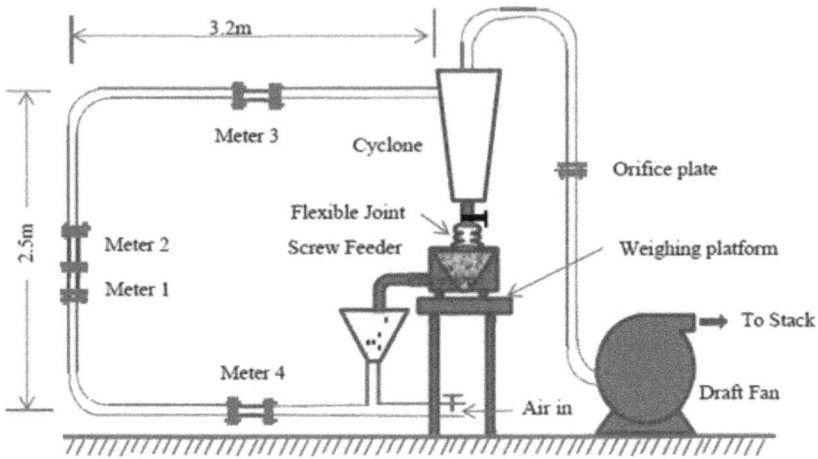

Figure 3. *Schematic of the pneumatic conveying rig.*

Fillite was used as test material due to its similar density to the density of coal and good handling ability. Original sample was analysed, and the mean particle size was around 175 µm. The same sample of about 3kg was reused for each fixed velocity under a given concentration for 8 to 11 runs (iterations), so that after each run, the mean size decreased due to particle attrition. For each different velocity, a new sample was taken from the same bag of the original Fillite material. The solids were discharged before they were re-used. A small amount of sample was collected after each test for future size distribution analysis.

Table 1. *Experimental parameters*

Material	Fillite
Size-d_{50} (µm)	175
Particle density (kg/m^3)	2000
Air/solids mass flow rate ratio	2 & 4
Number of runs	8 ~ 11

3. SIMULATION SETUP

The simulation technique used in this research was a coupled CFD and DEM approach[6]. Key to the coupling between the Computational Fluid Dynamics and Discrete Element Method is proper consideration of particle–fluid interaction forces. Typical particle–fluid interaction forces considered in past studies include the buoyancy force, pressure gradient force, drag force due to the particle motion resistance by stagnant fluid, as well as other unsteady forces such as virtual mass force, Basset force and lift forces. Following the approach proposed by Tsuji et al.[7,8] it was assumed that the motion of particles in the DEM is governed by the Newton's laws of motion and the pore fluid is continuous which can be described by locally averaged Navier–Stokes equation to be solved by the CFD[9]. The interactions between the fluid and the particles are modelled by exchange of drag force and buoyancy force only. The flowchart[10] in Figure 4 illustrated the principle of this simulation. Detailed formalisms governing the three aspects and numerical solution procedures are described as follows.

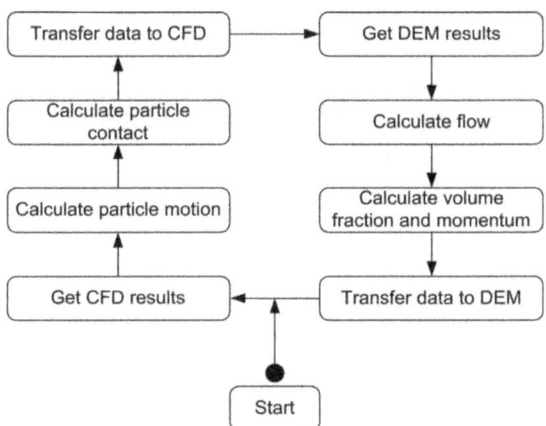

Figure 4. *Flowchart of the CFD-DEM simulation computation process.*

3.1 DEM analysis

The DEM code used for the modelling process in this paper was LIGGGHTS. For the particle-particle and particle-wall contacts, the Hertz-Mindlin contact model was selected.

$$\vec{F_n} = \vec{F}_{n,k} + \vec{F}_{n,d} = \left(k_{HM}\delta_n{}^{3/2} + c_{HM}\vec{v}_n\delta_n{}^{1/4}\vec{n}\right)\vec{n} \tag{1}$$

where k_{HM} is the normal stiffness; δ_n is the normal overlap, c_{HM} is the normal damping coefficient which is a function of the restitution coefficient C_R; $\vec{v}_n$ is the normal relative velocity; and $\vec{n}$ is the unit vector from the center of the colliding particle. The stiffness of the Hertz–Mindlin model is defined by:

$$k_{HM} = \frac{4}{3}E_{eff}\sqrt{\frac{D_{eff}}{2}} \tag{2}$$

where E_{eff} is the effective Young's modulus and D_{eff} is the particle diameter, calculated from Eq. 3

$$\frac{1}{D_{eff}} = \frac{1}{D_i} + \frac{1}{D_j} \tag{3}$$

$$\frac{1}{E_{eff}} = \frac{1-v_i{}^2}{E_i} + \frac{1-v_j{}^2}{E_j} \tag{4}$$

where v is the Poisson's ratio, and E is the Young's modulus.

Spherical particles were selected in the DEM modelling to replace the irregularly shaped particles which were computationally expensive. Nevertheless, to take the irregular shape effect of the particles into account, a rolling friction model was introduced. Generally, three types of rolling friction models were utilised, which was explicitly reviewed in the previous work[11]. This research selected a revised Type C rolling friction model which was achieved by Wensrich & Ketterfeld[12]. Detailed DEM simulation parameters were tabulated in Table 2.

Table 2. *DEM simulation parameters.*

Particle friction	0.5	Young's Modulus (Pa)	5×10^7
Rolling friction	0.2	Poisson ratio	0.3
Particle density (kg/m^3)	2000	Coefficient of restitution	0.3
Size (μm)	100,110,120,130,140 and 150	Mass (kg)	0.001

3.2   CFD modelling

The CFD code used in this simulation was OpenFOAM[13] with a transient flow solver 'Pisofoam' selected. The flow turbulence model used was standard k-epsilon which has been widely used in various CFD studies[14]. The drag force model adopted was the Koch-Hill model[15]. According to such a model, the drag force on a single particle was expressed as

$$\beta = \frac{3}{4}\frac{C_D\alpha_s\alpha_g\rho_g|\overrightarrow{u_g}-\overrightarrow{u_s}|}{d_p} \tag{5}$$

where C_D is the drag factor, α_g is the volume fraction of the gas phase, α_s is the particle volume fraction, ρ_g is the gas density, d_p is the particle diameter and $|\overrightarrow{u_g}-\overrightarrow{u_s}|$ is the absolute relative interracial velocity of the particles compared to the fluid.

The drag factor C_D is modelled as

$$C_D = 12\frac{\alpha_g{}^2}{Re_r}F \tag{6}$$

F is a dimensionless drag factor which correlates the drag to the Reynolds number and particle concentration. For this model, the Reynolds number Re_r is based on the radius of the particles rather than the diameter, which entails:

$$Re_r = \frac{\rho_g\alpha_g d_p|\overrightarrow{u_g}-\overrightarrow{u_s}|}{2u_g} \tag{7}$$

The dimensionless drag factor F is a piecewise function which is defined as:

$$F = 1 + \frac{3}{8}Re_r \quad \text{when} \quad \begin{cases} \alpha_s \leq 0.01 \text{ and} \\ Re_r \leq \dfrac{(F_2-1)}{\left(\dfrac{3}{8}-F_3\right)} \end{cases}$$

$$F = F_0 + F_1 Re_r^{\,2} \quad when \quad \begin{cases} \alpha_s > 0.01 \text{ and} \\ Re_r \leq \dfrac{F_3 + \sqrt{F_3^{\,2} - 4F_1(F_0 - F_2)}}{2F_1} \end{cases}$$

$$F = F_2 + F_3 Re_r \quad Otherwise \tag{8}$$

Model parameters such as F_0, F_1 F_2 and F_3 are defined in the following expressions:

$$F_0 = \begin{cases} (1-w)\left[\dfrac{1+3\sqrt{\frac{\alpha_s}{2}}+\left(\frac{135}{64}\right)\alpha_s \ln(\alpha_s)+17.14\alpha_s}{1+0.681\alpha_s-8.48\alpha_s^{\,2}+8.16\alpha_s^{\,3}}\right] + w\left[10\dfrac{\alpha_s}{(1-\alpha_s)^3}\right], & 0.01 < \alpha_s < 0.4 \\[4mm] 10\dfrac{\alpha_s}{(1-\alpha_s)^3}, & \alpha_s \geq 0.4 \end{cases}$$

$$\tag{9}$$

$$F_1 = \begin{cases} \dfrac{\sqrt{\frac{2}{\alpha_s}}}{40} & , 0.01 < \alpha_s < 0.1 \\[4mm] 0.11 + 0.00051e^{(11.6\alpha_s)} & , \quad \alpha_g > 0.1 \end{cases} \tag{10}$$

$$F_2 = \begin{cases} (1-w)\left[\dfrac{1+3\sqrt{\frac{\alpha_s}{2}}+\left(\frac{135}{64}\right)\alpha_s \ln(\alpha_s)+17.89\alpha_s}{1+0.681\alpha_s-11.03\alpha_s^{\,2}+15.41\alpha_s^{\,3}}\right] + w\left[10\dfrac{\alpha_s}{(1-\alpha_s)^3}\right], & \alpha_s < 0.4 \\[4mm] 10\dfrac{\alpha_s}{(1-\alpha_s)^3}, & \alpha_s \geq 0.4 \end{cases}$$

$$\tag{11}$$

$$F_3 = \begin{cases} 0.9351\alpha_s + 0.03667 & , \alpha_s < 0.0953 \\[2mm] 0.0673 + 0.212\alpha_s + \dfrac{0.0232}{(1-\alpha_s)^5} & , \alpha_s \geq 0.0953 \end{cases} \tag{12}$$

$$w = e^{(-10(0.4-\alpha_s)/\alpha_s)} \tag{13}$$

The gas phase hydrodynamics are calculated in three dimensions from the Navier-Stroke equations for the fluid in presence of a granular phase:

$$\frac{\partial \alpha_g \rho_g}{\partial t} + \nabla \cdot \left(\alpha_g \rho_g \overrightarrow{u_g}\right) = 0 \tag{14}$$

$$\frac{\partial(\alpha_g \rho_g \overrightarrow{u_g})}{\partial t} + \nabla \cdot \left(\alpha_g \rho_g \overrightarrow{u_g} \overrightarrow{u_g}\right) = -\alpha_g \nabla p - \beta\left(\overrightarrow{u_g} - \overrightarrow{u_s}\right) + \nabla \cdot \left(\alpha_g \overrightarrow{\tau}\right) + \alpha_g \rho_g \overrightarrow{g} \tag{15}$$

The drag from the Eq. 5 was incorporated into Eq.15. Air was selected as the gas phase. Standard properties of air at 25 ^{0}C were selected.

4. PARTICLE ATTRITION RESULTS

After conducting series of experiments with the above discussed setting, mean particle size of the degraded sample at different run numbers was initially analysed. Figure 5. depicts particle degradation at different velocities under air to solids ratio of 4. As expected, results indicated that the mean particle size decreased with the increasing number of runs for this material. This was due to the particle attrition which effectively broke large sized particles into smaller particles. It can also be noted that the particle size reduction was severer at higher flow velocities which was because the particle-particle/particle-wall collision were more vigorous in these scenarios.

There were errors in mean size analysis, which can be clearly seen for v=12m/s where the mean size at 8th run was greater than the original size. The errors were due to sample collection, the shaker time setting and other error sources. These errors became intolerable when the actual size change was small. Nonetheless, Figure 5 does show that the velocity is key parameter for particle degradation.

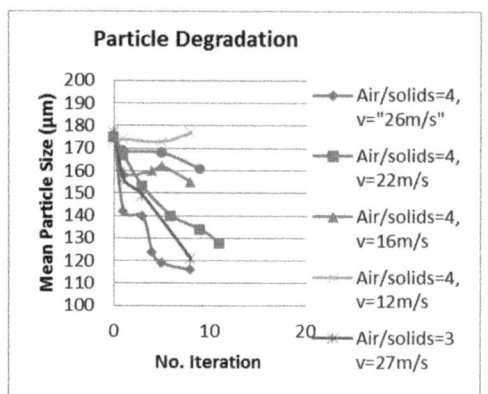 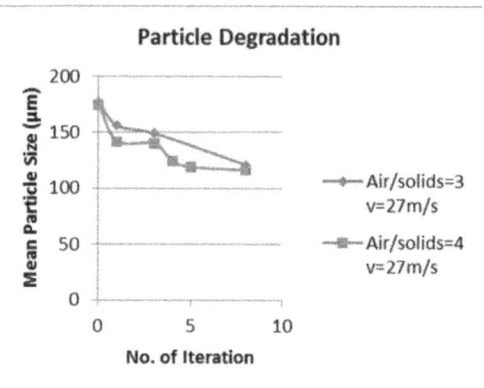

Figure 5. *Particle attrition results; (a) particle degradation with run number; (b) effect of air/solids ratio on particle attrition.*

From Figure 5(b), it seems that the leaner phase (higher air to solids ratio) could result in severer particle attrition.

5. CORRELATION OF SENSOR OUTPUT VARIATION AND PARTICLE ATTRTION

The response of a dynamic electrostatic meter depends on the combination of flow turbulence level and total charge carried by particles. When particles break into smaller particles, the total charge carried by the particles increases and the particle turbulence level decreases. The rise or fall of the signal trend depends on the shift of the dominant effect.

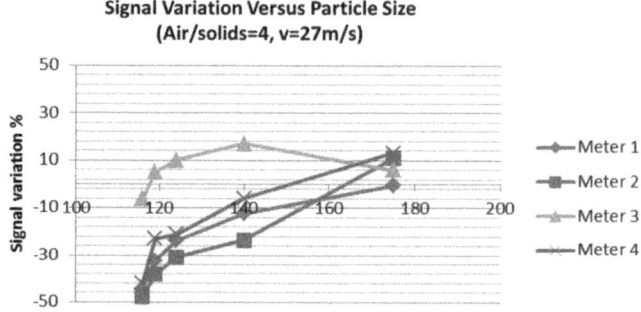

Figure 6. *Responses of Electrostatic Meters to Particle Attrition.*

This can be seen from Figure 6 where the responses of the four meters (the locations of them can be found in Figure 3) are illustrated. Overall, the signal level fell when particles became smaller. However the output of Meter 3 increased when the mean particle size dropped from 175μm to 140μm, indicating over the range of 175-140 μm, it was the increase in total charge dominated the output at this location, probably the turbulence level at this location fell slowly than at other locations. As previously discussed, the signal fluctuation level indicated the turbulence degree in the flow and the total charge carried by particles.

4.3 Simulation Results

The simulations were conducted based on the dimensions and layout of the Teesside pneumatic conveying rig as shown in Figure 7. Additionally, the result for particle size of 100 μm at different simulation time interval was shown in Figure 8. As can be seen from the figure, the air-solids flow exhibited similar phenomenon as observed in common pneumatic conveying experiments. After entering into the pipe, the material was firstly accelerated by the air flow. Then, it was re-accelerated after each bend in the pipe geometry. During this process, particle migration, stratification and unsteady flow characteristics were all observed.

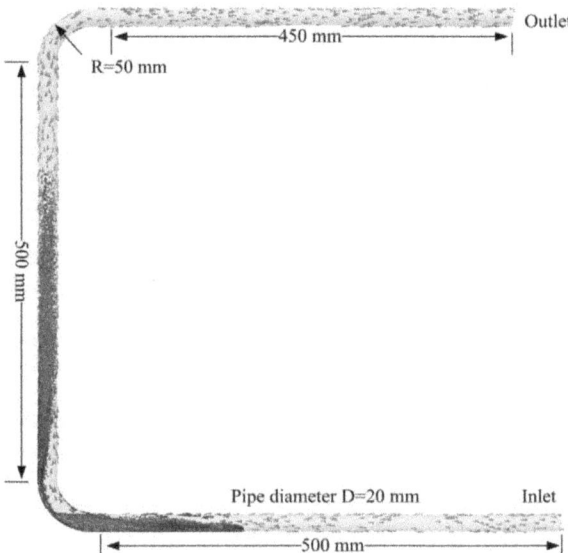

Figure 7. *Simulation results of material with particle size of 100 μm.*

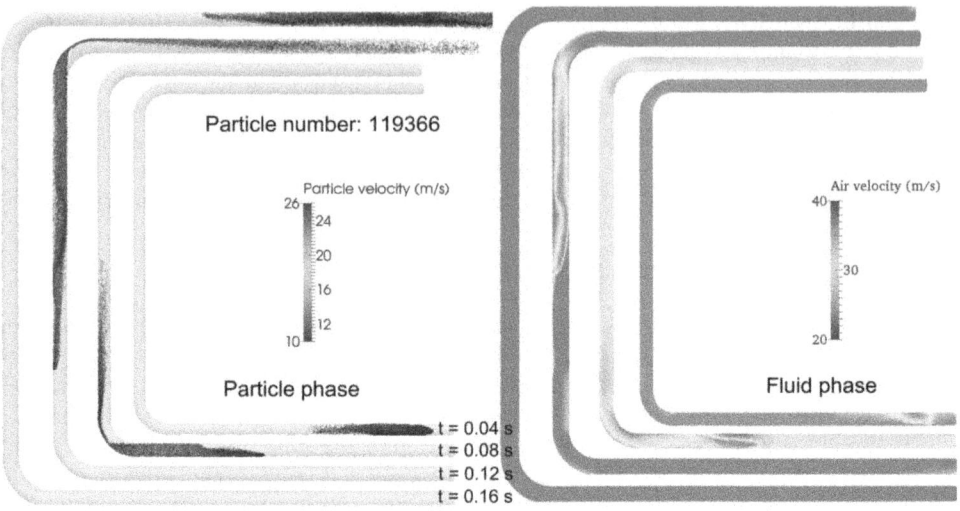

Figure 8. *Simulation results for the particle phase (100μm) and fluid phase at different time intervals.*

In principle, flows mechanism of materials with different particle size will vary to a large extent. This can be observed by comparing the internal flows at a specific simulation time interval. Internal flows at 0.07s in all 6 simulations were presented in Figure 9. As can be seen from the figure, air-particle flows exhibited distinct interactions in different simulation scenarios.

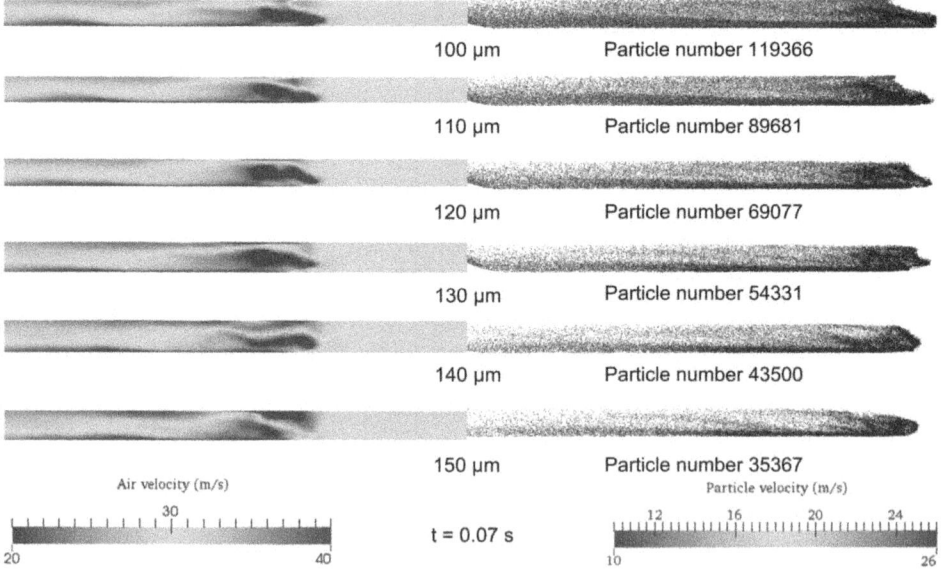

Figure 9. *Simulation results of each particle size at 0.07 time interval.*

Additionally, particle velocity components ($\overrightarrow{u_s} = \overrightarrow{u_{sx}} + \overrightarrow{u_{sy}} + \overrightarrow{u_{sz}}$) at a specific time interval in the simulation were obtained. Subsequently, depending on the position of the particle in the pipe geometry, the velocity components were then transformed into axial and radial velocity for further turbulence investigation. Average values of the two velocity components in each simulation after passing the Meter 4 position were extracted from the simulation results and subsequently tabulated in Table 3.

Table 3. *Mean particle velocity values when particle passing Meter 4.*

Particle size (μm)	Mean axial velocity (m/s)	Mean radial velocity (m/s)
100	10.39	0.42
110	10.04	0.41
120	9.22	0.39
130	8.87	0.38
140	9.35	0.40
150	9.65	0.42

In order to compare the turbulence level in the simulation with the experiment, a simple turbulence ratio was proposed in the following expression,

$$T_r = \left| \frac{v_{radial}}{v_{axial}} \right| \tag{16}$$

where v_{radial} is the radial velocity and v_{axial} velocity of each particle, respectively. Such a definition indicates that there is more turbulence in the flow when the T_r ratio is higher.

Based on this definition, the turbulence ratio for each particle in all simulation scenarios were calculated and averaged. As showed in Figure 10, such a turbulence ratio value was observed to increase with the increment of the particle size. Consequently, this implied that the turbulence is severer in larger sized particles than smaller sized particles, which verified the results obtained from the electrostatic sensors in the experiments.

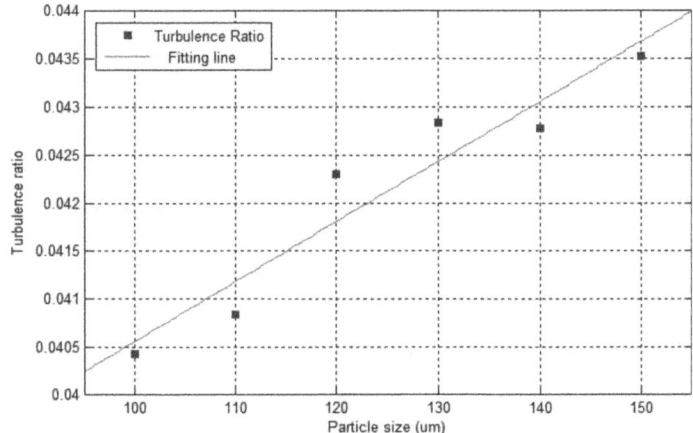

Fig. 10. *Correlation between the turbulence ratio and particle size in all simulations.*

6. CONCLUSION

The research presented the study on the influence of the particle size on the turbulence of gas-solids flows. Due to the particle attrition, the mean particle size will decrease during the pneumatic operation. Both experimental and numerical methods were utilised to conduct the study. In terms of the experiment, 4 electrostatic sensors were installed in a pilot pneumatic conveying system to monitor the flow turbulence levels. 8 to 11 runs were carried out using the Fillite material. Mean particle size was measured from the sample material collected after each run. Experimental results indicated that the particle size was decreasing along with the run number, and the corresponding electrostatic sensor signal was also in reduction which embodied that the turbulence in the flow in decreasing.

Additionally, a CFD-DEM coupled simulation technique was also adopted to study this phenomenon. 6 simulations with various particle size (100, 110, 120, 130, 140 and 150 μm) were performed. Particle axial and radial velocity components during the conveying process were analysed. A turbulence ratio of the flow was defined as the ratio of particle radial velocity to the axial velocity. The turbulence ratio is proportional to the turbulence level within the flow. Simulation results showed that the turbulence ratio increased with an increasing particle. Consequently, this observation verified the experimental results.

Reference

1   J. (Jianyong) Zhang, J. M. (John) Coulthard, and B. (Brian) Armstrong, "Lean phase solids control using pinch valves and mass flowmeters for pulverised fuel," Jan-2003.
2   C. Xu, J. Zhang, D. Yang, B. Zhou, G. Tang, and S. Wang, "Dense-phase pneumatically conveyed coal particle velocity measurement using electrostatic probes and cross correlation algorithm," *J. Phys.: Conf. Ser.*, vol. 147, no. 1, p. 012004, Feb. 2009.
3   B. Zhou and J. Zhang, "Potential measurement in ECT system," *Journal of Electrostatics*, vol. 67, no. 1, pp. 27–36, Feb. 2009.
4   K. Saleh and A. Aghili, "The spatio-temporal evolution of tribo-electric charge transfer during the pneumatic conveying of powders: Modelling and experimental validation," *Chemical Engineering Science*, vol. 68, no. 1, pp. 120–131, Jan. 2012.
5   J. Zhang, J. Coulthard, R. Cheng, and R. Keech, "Measuring Pulverised Fuel: Using Electrostatic Meters," *Measurement and Control*, vol. 42, no. 3, pp. 87–90, Apr. 2009.
6   C. Goniva, C. Kloss, and S. Pirker, "Towards fast parallel CFD-DEM: An open-source perspective," in *Proc. Open Source CFD International Conference, Barcelona*, 2009.
7   Y. Tsuji, T. Tanaka, and T. Ishida, "Lagrangian numerical simulation of plug flow of cohesionless particles in a horizontal pipe," *Powder technology*, vol. 71, no. 3, pp. 239–250, 1992.
8   Y. Tsuji, T. Kawaguchi, and T. Tanaka, "Discrete particle simulation of two-dimensional fluidized bed," *Powder Technology*, vol. 77, no. 1, pp. 79–87, Oct. 1993.
9   T. B. Anderson and R. Jackson, "Fluid mechanical description of fluidized beds. Equations of motion," *Industrial & Engineering Chemistry Fundamentals*, vol. 6, no. 4, pp. 527–539, 1967.
10 C. Goniva, C. Kloss, A. Hager, and S. Pirker, "An open source CFD-DEM perspective," in *Proceedings of OpenFOAM workshop Gothenburg, Sweden*, 2010.
11 J. Ai, J.-F. Chen, J. M. Rotter, and J. Y. Ooi, "Assessment of rolling resistance models in discrete element simulations," *Powder Technology*, vol. 206, no. 3, pp. 269–282, Jan. 2011.

12 C. M. Wensrich and A. Katterfeld, "Rolling friction as a technique for modelling particle shape in DEM," *Powder Technology*, vol. 217, pp. 409–417, 2012.

13 H. Jasak, A. Jemcov, and Z. Tukovic, "OpenFOAM: A C++ library for complex physics simulations," in *International Workshop on Coupled Methods in Numerical Dynamics, IUC, Dubrovnik, Croatia*, 2007, pp. 1–20.

14 V. C. Patel, W. Rodi, and G. Scheuerer, "Turbulence models for near-wall and low Reynolds number flows - A review," *AIAA Journal*, vol. 23, no. 9, pp. 1308–1319, Sep. 1985.

15 R. J. Hill, D. L. Koch, and A. J. Ladd, "Moderate-Reynolds-number flows in ordered and random arrays of spheres," *Journal of Fluid Mechanics*, vol. 448, no. 2, pp. 243–278, 2001.

IMAGE VISUALIZATION OF MICRO-STRUCTURES IN THE ENTRAINMENT OF JET FLOW BY USING SFSEI METHOD

Z H Zhu, W Zhou, X S Cai

Institute of Particle and Two-phase Flow Measurement, University of Shanghai for Science and Technology, Shanghai, 200093

1   INTRODUCTION

The Entrainment is the key to study the turbulent problem of jet flow[1]. The entrainment of ambient fluid along the flow path increased the flow rate and decreased the flow velocity. The entrainment phenomenon was first introduced by Sir Geoffrey Taylor, and wider expressed through the review lecture by Batchelor in 1954[2]. From then on, a large number of theoretical and experimental researches were finished[3-5].

Most experimental methods, such as Laser-induced Fluorescence (LIF) and Particle Image Velocimetry (PIV), were mainly used to recover the macro structures. Few measurements could reach micron sized level[6]. Depended on the development of computer technology, the numerical simulation has been a good application to jet flow. The size of the grids could reach micron sized level, and the number of the grids could reach hundreds of millions[7]. In order to make the comparison of numerical simulation results, the experimental study of micro structures of entrainment is necessary.

The Single Frame and Single Exposure Imaging (SFSEI) is a method suitable to visualize the micro structures of entrainment. In this paper, the micron sized vortexes of the entrainment for jet flow were recorded by using SFSEI method. The obtained structures of entrainment were as sophisticated as the results of numerical simulation. Otherwise, the generating process of entrainment was continuous recorded to analysis the mechanism of entrainment generating.

2.   OULINES OF JET FLOW EXPERIMENT

2.1 Theory of SFSEI method

SFSEI method is different to PIV method, captured the trajectories of particles with single frame in a relative longer exposure time. The velocity of particle in flow calculated as follow.

$$V_p = \frac{S_p}{\Delta t} \qquad (1)$$

Where V_p is the velocity of particle, S_p is the length of the particle's trajectory and Δt is the exposure time.

The advantages of this method can be concluded as: 1. visualize the flow field directly: the streamlines were shown directly in the picture of the trajectories of particles taken by single frame; 2. able to study micro-structures: the size of pixels in CMOS sensor, in this work, was $5.3 \times 5.3 \mu m^2$, and the diameter of the smallest vortex in entrainment reached 21.2 μm^2; 3. low equipment requirements: the industrial camera and the continuous laser used in this work, were much easier to realize online measurement and lower price than PIV's high-speed camera and pulsed laser.

The disadvantage is that the trajectory of particle is difficult to show the flow direction. The flow direction can only be obtained by the knowledge of Hydromechanics and the experience. However, the jet flow is a classical problem in Fluid Dynamics, the flow direction is easy to judge.

2.2 Experimental setup and settings

The experimental setup is shown in Figure.1. The main components were Water Circulation System, Rotameter, Jet Flow Region, SFSEI System. Water Circulation System can provide stable flow rate, and the flow rate is changeable. The function of Rotameter is to limit the flow and just estimate the flow rate. The real flow rate is calculated by SFSEI method. The Jet nozzle is a straight-wall nozzle which diameter is 5mm.

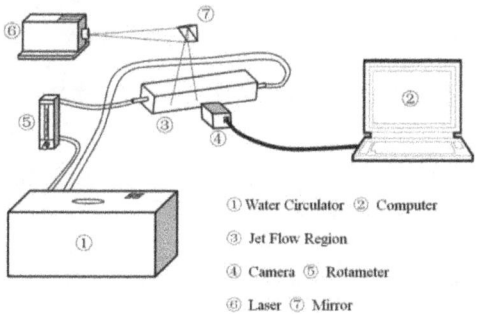

① Water Circulator ② Computer
③ Jet Flow Region
④ Camera ⑤ Rotameter
⑥ Laser ⑦ Mirror

Figure 1. *Experimental setup.*

As shown in Figure.1, the sheet light produced by continuous laser and spherical lens, which is thinner than 1 mm. The particles (Al_2O_3) were lighted by the sheet light, and the trajectories of particles were caught by the CMOS camera. At this work, the continuous laser's wavelength was 532nm, the Reynolds numbers was 1241.3, the camera's exposure time was 5ms, the fps was 50, the observation size was $6.36 \times 3.97 mm^2$, the spatial resolution was $5.3 \times 5.3 \mu m^2$.

3. RESULTS AND CONCLUSIONS

The generating process of entrainment is shown in fig.2. The process was divided into four stages: 1. the start of jet flow, 2. the huge vortex passed, 3. slight entrainment, 4. obvious entrainment.

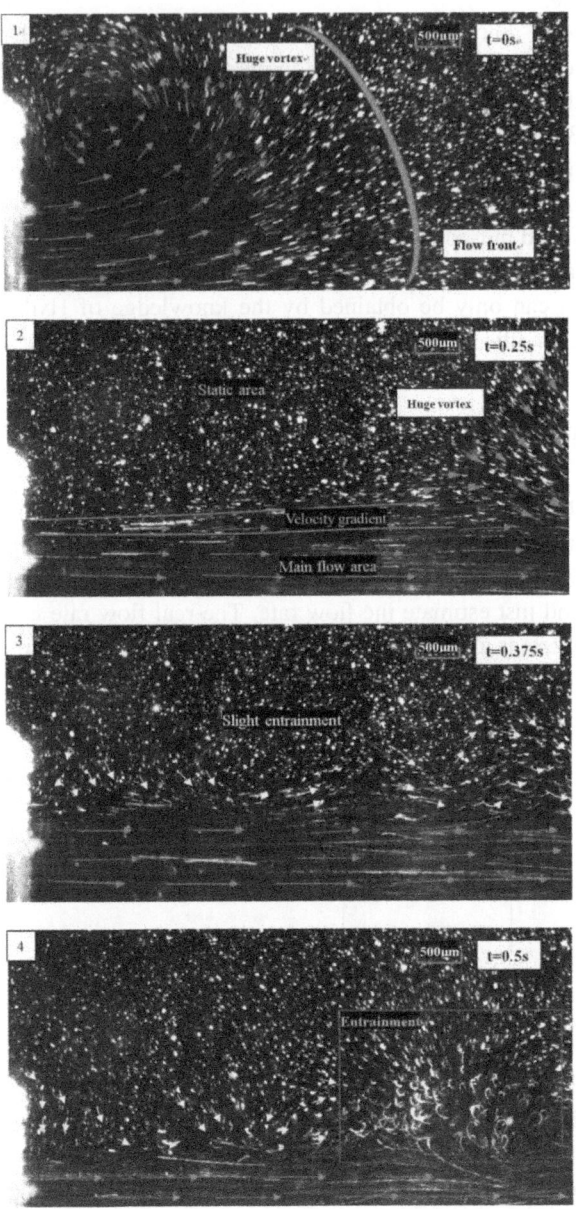

Figure 2. *The generating process of entrainment when Re=1241.3*

At the start of flow, when t=0s, there was a huge vortex generated by the impetus of flow. The huge vortex moved forward with the flow front. At stage 2, there was significant instantaneous velocity gradient between 'Static area' and 'Main flow area'. So the stage 2 was unstable, as the stage 3 shown, slight entrainment appeared near the surface of 'Main flow area'. The entrainment grew up as time went on. When t=0.5s, the obvious structures of entrainment generated. As the Figure.3 shown, the refined structures of entrainment part were presented.

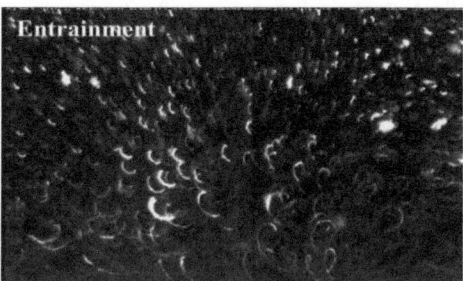

Figure 3. *The amplified structures of entrainment part*

The entrainment part was consisted with a large number of small and independent vortexes. As shown in Figure.3, because of the velocity gradient, the area close to main flow was filled with vortexes which are larger than those in the area away from main flow. The diameter of the largest vortex reached 478.7μm. And the diameter of smallest vortex around the periphery of entrainment reached 21.2μm. Even though the Re number is low, which is 1241.3, the structures of entrainment showed in Figure.3 revealed the obvious turbulence. The small turbulent structures were tiny and complicated that difficult to be acquired by PIV. The velocity vector field of the entrainment part was shown in Figure.4.

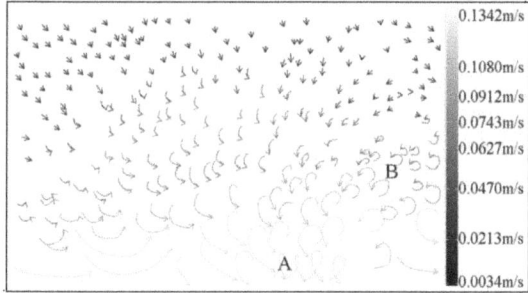

Figure 4. *The velocity vector field of entrainment part*

As the velocity vector field showed, the fluid outside the main flow converged on point A along with the velocity increased. When the velocity reached 0.47m/s, the curved trajectories appeared. And the trajectories in area B, under the effect of double entrainment of point A and downstream, were turned to be dramatic rolls. The refined velocity vector field showed in Figure.4 was helpful to understand the mechanism of the entrainment in jet flow.

4. CONCLUSIONS

The micro structures of entrainment of jet flow were visualized by using Single Frame and Single Exposure Imaging (SFSEI) method. The conclusions can be drawn as follows.

1. SFSEI method is suitable for visualization of micro-structures in multi-phase flow.

2. The micro structures of entrainment were consisted with independent small vortexes. The scales of these vortexes reached micron sized level which the sized level of numerical simulation. And the smallest vortex's diameter was 21.2μm.

3. On the micron-scale, the structures of entrainment were more complicated and turbulent.

ACKNOWLEDGEMENT
The authors acknowledge the support from the National Natural Science Foundation of China (Grant No. 51206112 and Grant No. 51327803); Shanghai Natural Science Foundation (Grant No. 12ZR1446900).

References

1. J. S. Turner, *J. Fluid Mech.* 1986, 173, 431.
2. G. K. Batchelor, *Q. J. R. Meteorol. Soc.* 1954, 339.
3. M. Choi, H. S. Yoo, G. Yang, et al. *Int. J. Heat Mass Transfer.* 2000, 43, 1811.
4. D. B. Li, J. R. Fan, K. F. Cen, *Fuel.* 2012, 102, 470.
5. C. Eichler, T. Sattelmayer, *Exp Fluids.* 2012, 52, 347.
6. D. Liepmann and M. Gharib, *J.Fluid Mech.* 1992, 245, 643.
7. P. He, Z. Zhou, et al. *J. Visualization*, 2009, 12, 286.

Subject Index